A MATLAB® Manual for

ENGINEERING MECHANICS

STATICS

COMPUTATIONAL EDITION
SI EDITION

Robert W. Soutas-Little
Michigan State University

Daniel J. Inman
Virginia Tech

Daniel S. Balint
Imperial College London

CENGAGE Learning™

Australia • Brazil • Japan • Korea • Mexico • Singapore • Spain • United Kingdom • United States

CENGAGE Learning™

A MATLAB® Manual for Engineering Mechanics: Statics, Computational Edition, SI Edition
Robert W. Soutas-Little, Daniel J. Inman, and Daniel S. Balint

Director, Global Engineering Program: Chris Carson
Senior Developmental Editor: Hilda Gowans
Permissions: Vicki Gould
Production Service: RPK Editorial Services, Inc.
Copy Editor: Pat Daly
Proofreader: Erin Wagner
Indexer: Shelly Gerger-Knechtl
Creative Director: Angela Cluer
Text Designer: RPK Editorial Services
Cover Designer: Andrew Adams
Cover Image: Zac Macauley/The Image Bank/ Getty Images
Compositor: International Typesetting and Composition
Printer: Thomson/West

Library of Congress Control Number: 2008922462
ISBN-13: 978-0-495-43813-7
ISBN-10: 0-495-43813-8

Cengage Learning
1120 Birchmount Road
Toronto ON M1K 5G4 Canada

Printed in the United States of America
1 2 3 4 5 6 7 11 10 09 08

Contents

Introduction

This supplement to *Engineering Mechanics: Statics—Computational Edition* by Soutas-Little, Inman and Balint, will provide all the necessary instructions to use recent versions of MATLAB® software, and in particular the *Student Version* of MATLAB available from the web at http://www.mathworks.com/academia/student version/, to aid in solving the homework problems and working through the sample problems. It should be noted that before any attempt is made to use this computational software, the problem must be correctly modeled, that is, a free-body diagram constructed and the correct equations of equilibrium formulated. Computational software only reduces the numerical burden, makes it possible to conceptualize the solution through plots, and facilitates parametric studies which enable design and physical principles to be examined. MATLAB is a registered trademark of The Mathworks, Inc., 3 Apple Hill Drive, Natick, Massachusetts, 01760. Both the *Student Version* and professional versions are available from The Mathworks. Many institutions have a site license for student access and the *Student Version* is often available in campus bookstores. The web address listed above allows a free download of the manual for MATLAB, if more detailed instructions are of interest. The instructions for the student and professional versions are the same. Just the prompt is different as described below.

A large number of universities have fantastic web sites introducing the use of MATLAB. If you run in to trouble understanding any aspect of the code the answer can often be found by simply typing "matlab" into Google which will reveal a host of excellent sources. In addition, the Help file in MATLAB is also very useful. The Mathworks web site is the best source for using the code as The Mathworks continually updates their software and their web site.

This supplement is intended to guide the reader through the use of MATLAB for solving statics problems. As such, you are encouraged to open your version of MATLAB and try the various steps described in the following as you read through it. It is keyed heavily to the accompanying text and works through many of the sample problems in detail, solving the sample problems from the text using MATLAB for each chapter. It is suggested that you work through the supplement with MATLAB open on your computer until you get to Chapter 2 of the *Statics* text. Then refer to the supplement as needed in solving the homework and studying the text. This supplement is written in a style consistent with the *Student Version* as used in most MATLAB documentation and is keyed to the *Statics* text material. While this supplement suggest ways to use MATLAB to enhance your understanding of Statics and teach you efficient computational skills, you may browse through the *Student Version* or professional version manuals and think of your own usage of MATLAB to solve statics problems.

Wherever possible, the input to, and the output from MATLAB has been copied and pasted directly into this text. The fonts however, have been modified to make clear exactly what the reader should type in order to reproduce the results discussed in this manual. There are three fonts used here. The font for the base text which provides information and discussion, the font to indicate that it is MATLAB generated: `EDU>`, and the font indicates information that is entered into MATLAB by the user: `x=[35;45;10]`. In some cases the spacing has been changed to "fit" the page and to avoid large empty spaces. Note the prompt for the student version is `EDU>`, while the prompt for the professional version is `>>`.

This supplement consists of 10 sections. The first section is an introduction to using MATLAB, concluding with a sample statics problem and can be studied while reading Chapter 1 of the Statics text. Nine more sections follow this, one for each of the chapters 2 through 10 of the companion Statics text (*Engineering Mechanics: Statics—Computational Edition*, by Soutas-Little, Inman and Balint). Each of these remaining sections presents MATLAB solutions for the Sample Problems given in the Statics text.

1

Using MATLAB

Numerical Calculation

In using MATLAB for the first time, it is useful to think of the package as a calculator. (If you are familiar with MATLAB, you may wish to skip this section.) In fact, it is worth while to load the package and start using it to make some very simple calculations (add, multiply, etc.) to become comfortable with its format, much as you would with a newly purchased calculator. In MATLAB, numbers are entered and assigned a value by using an equals sign. When you open MATLAB, the following prompt will appear:

```
EDU>
```

This prompt is displayed inside a window called the *Command* window. To assign the symbol "a" the value 0.24, just type `a = 0.24` after the prompt. Your screen will then show the following:

```
EDU> a=0.24
```

In MATLAB, there is no need to worry about spaces as they are usually ignored (except in the case of entering matrix data). As soon as you hit the return key, MATLAB returns the result of your entry. In this case, you will see the following on your screen:

```
EDU>a=0.24
a =
        0.24
```

If you do not want MATLAB to write out the result onto the screen you may suppress this feature by adding a semicolon at the end of the statement (i.e., in this example, type EDU>a=0.24;). If you type any legitimate list of operations, they will be computed in this step. For example, we might have

```
EDU>c=3*a^2
c =
     0.1728
```

If MATLAB cannot recognize a symbol or command, it will give you an error message with some hint about what might be wrong.

MATLAB deals with numbers in lists, called vectors, and in arrays, called matrices. Even a scalar is treated as a 1×1 matrix. Thus, if the symbol u is to denote a row vector of, say, the coordinates of a particle, then type

```
EDU> u=[1 -1 2]
u=
       1    -1    2
```

Here columns are separated by a space (a notable exception to the general rule that spaces do not matter). If a column vector is desired, separate the entries by semicolons and type

```
EDU> u=[1;-1;2]
u=
       1
      -1
       2
```

The semicolon is used to denote a new row. Alternately, you could define v=u' to get the column vector from the row vector u. The apostrophe denotes the transpose of the vector or matrix that precedes it. These commands and associated manipulations are extremely useful in statics, as are all the vector manipulations (dot product, cross product, addition, and subtraction) that are also defined in MATLAB.

The MATLAB code is based on using arrays of numbers. So it also uses this notion of a vector to keep a record of variables that have a range of discrete values. For instance, in the example to follow, the constant L is defined to be 10 (meters), and the variable x is defined to start at 0 and continue to increment by 0.1 until the final value of x is 10, or L. Rather then entering the string of values that x takes on in a list as was just done, the list can be generated automatically by typing

```
EDU>x=(0:0.1:10);
```

Here the first number is the starting value of x, the second number is the increment, and the last number is the ending value. These entries are separated by colons. For example, if the variable x was to begin at 2 and have an increment of 0.1 and a final value of 8, x=(2: 0.1:8) would be entered. Note that the semicolon at the end of the line suppresses the display of the value of x. If you prefer to generate the list by specifying the number of values in x rather than specifying the increment, type

```
EDU>x=linspace(0,10,101);
```

This produces the same list of values of x as does the previous line of code.

Now that you know how to enter constants and variables and how to generate ranges of values of variables, the next step is to treat these values as the domain of a function. MATLAB uses most of the standard math functions (see the inside back cover for a list) and operates on a list of any or all of these functions in a single command. For example, to evaluate sin(x), just type

```
EDU> y=sin(x);
```

The statement assigns the row vector y the various values of sin(x) for each discrete value of x contained in the list for x. By this command, the MATLAB code assumes that you want to calculate the sine of each element in the vector x and place that value in an associated row vector y. However, a little care must be taken in multiplying and adding variables that represent vectors. If you need to multiply a variable `x` consisting of 101 elements by another vector variable `y`, it must be of the same number of elements (101). For example, to multiply each element of `x` by each element of the same-size vector `y`, type `x.*y`. The period before the multiplication operator signifies an element-by-element multiplication. In particular, if the vectors `x = [1 0]` and `y = [-1 2]`, then typing `x.*y` returns `ans = -1 0`, which is a row vector. This is not a vector product used in mechanics, but the operation does modify the procedure for manipulation of variables in digital form using MATLAB.

Significant Figures

MATLAB computes all numbers to a high degree of numerical accuracy. You can choose how to display these numbers in terms of how many digits (5 or 16), but none of this is a reflection of the number of significant digits in your "answer." Thus, it is very important to interpret the results of any computer output using the rules for significant figures and common sense. For example, if you have MATLAB compute the average number of people driving by your home on a given day by using a series of MATLAB commands and the number returned is 50.533 or 50.53333333333334, common sense tells you that this is either 50 or 51 people. The point to remember is that the numbers returned from your commands must be further analyzed to determine their meaning.

Symbolic Calculations

MATLAB also allows you to make certain evaluations symbolically. For instance, you can write a function and differentiate it symbolically. In order to let MATLAB know that a variable is to be treated symbolically, rather than numerically (as was previously done), the variable is placed between single quotation marks. For example, `x` denotes a numerical-valued quantity, while `'x'` denotes a symbolic variable. Also, symbolic objects can be created from numeric ones by using the sym function. For example, `x=sym('x')` creates a symbolic variable `x`. Once a variable is defined as symbolic, it can be used in mathematical expressions the same way that numerical variables can, except that the corresponding evaluation or computation is made symbolically. For example, the following code differentiates the sine function:

```
EDU>x=sym('x');    % this line creates a symbolic variable x
EDU>diff(sin(x))   % this line differentiates sin(x) with respect to x
ans=
      cos(x)
```

Note that the % symbol denotes that what follows is a comment and not part of the code. Also, MATLAB uses the term "`ans =`" to denote that the answer follows unless you have assigned the calculation another name. More details on using the symbolic aspects of MATLAB appear later in the text.

Saving Files

It is often desirable to be able to repeat the same calculation with different values. This allows you to perform "what if" studies and aids in design. In the foregoing commands, we have been working in the command window, which computes every time we hit the return key and has no ability to go back in our list of formulas to fix the commands if we find that we do not like our answer or if we cause an error. It often happens in engineering that, after a long list of formulas, we look at our answer and find that it is unreasonable. For instance, if we use a string of formulas and calculations to compute the weight of a car, and our answer turns out to be five pounds, we know we made an error somewhere in our calculations. It would be useful to be able to go back and find the line that needs to be corrected, fix it, and execute the code again without having to retype (or copy) the entire group of commands. You can do this by saving the file with all your commands as an "m-file," also called a *script* file. This special kind of file (described in paragraphs that follow) is essentially a script that MATLAB reads to perform the calculations you want. The names of such files end in ".m"; hence, they are also called m-files (not as exciting as *X-files*).

To create a list of formulas into an m-file, choose **New** from the file menu and select **m-file**. This will display a text editor window, in which you can now enter your commands. Enter

```
% this m-file converts inches to meters following Sample 1.2
D=15*2.540*0.001
```

into the window and save the file, naming it "samp1pt2.m". Be careful not to use a dash or period in naming the file, except, of course to end the name using .m. Now if you type the name of the file after the prompt (without the .m part), MATLAB will return the answer:

```
EDU>samp1pt2
D =
      0.0381
```

The difference between typing formulas in directly and using the m-file approach is that you can now go back and change the m-file if need be, without retyping the entire line. Suppose, then, that you would like the diameter in meters of a 14-in. tire. Just open samp1pt2.m, make the change, and run it again to get the new value (D = 0.03556). Effectively, the m-file allows you to edit your commands in a more-or-less spreadsheet fashion so that you can make changes to your code without reentering every command. This is not so important in the simple example used here, but it will become important as your knowledge of mechanics increases and the level of calculation becomes more involved. M-files are the feature that allows you to perform "what if" studies and to interrogate your answers to complex engineering problems. Often, long entries of data are subject to typing errors, and any statics problem is subject to intellectual errors (OK, mistakes). Such errors can frequently be found by changing a few numbers to achieve a new solution, which may or may not make sense. For example, sometimes setting some parameters to zero will reduce the problem to a simple one whose analytical solution is known to you. Then a comparison of the well-known analytical solution with your numerical solution will either give you confidence in your computer result or send you looking for an error in your code or in the logic of the mechanics behind it. This is an extremely important aspect of engineering and should become a routine part of your problem-solving technique.

Defining a Function

While MATLAB has a number of useful mathematical functions defined in its library (see the inside back cover), it is often very useful to be able to define your own functions. This can be done by creating a special m-file and using the existing functions in MATLAB. Consider the case of a load on a beam that is defined by the function $w(x) = 10\,(x - x\sin(x/10))$ in units of force per unit length (say, N/m). To define this as a function in MATLAB, create an m-file with the name of the function, using the already defined functions from the table. For example, suppose we decide to call this function "wload". Then create an m-file (select **New** and **M-File** from your menu) using this same name, with the first line indicating that it is a function of x as follows:

```
function y = wload(x)
                                    % this file defines the function wload as a function of x
y = 10*(x-x.*sin(1/10*x));
```

Give this file the name wload.m. (It is important that the function name in the file and the name of the file are the same. In fact, when you save the file, MATLAB will try to name it correctly for you.) Note that the first line must start with *function*, and in building the formula for y, the multiplication ". *" is used instead of just "*", because x and sin(x) are both vectors of numbers. The "*" by itself works for multiplying two scalars together (1 X 1 matrices) and when multiplying a scalar times a vector. If you need an element-by-element multiplcation or division use the dot notation (.* and ./, respecitvely). With this special function defined and saved, you can now call up the value wload(3.5) and get the value of the function when $x = 3.5$, rather then an error. If you type wload(3.5) without defining it as a function, MATLAB thinks you are looking for an entry in the vector wload in the 3.5th location, which does not exist. (Vectors have only integer locations.) The following code illustrates this use of the user-defined function wload:

```
EDU>wload(3.5)
ans =
       22.9986
```

Graphing

Next, consider using MATLAB to plot a graph. The ability to plot is the key to explaining your results to yourself and others, and it is also a great problem-solving technique. For instance, plotting complex expressions will facilitate the solution of difficult algebraic equations. MATLAB can create *x*-*y* plots, three-dimensional line plots, surface plots, polar plots, and vector plots. As an example, consider plotting a simple function, say sin(x). First the domain of the function (i.e., the values that x will take on) must be defined, as was previously done. Then the values of sin(x) must be determined and stored in the vector y (by using `y=sin(x)`, as defined earlier). Once all this is established, a simple plot command is utilized. Note the use of the semicolons to suppress displaying x and y, as it is only the plot we are interested in. The relevant code is

```
EDU>x=(0:0.1:10);   % defines the range of values of x
EDU>y=sin(x);       % defines the corresponding values of y
EDU>plot(x,y)       % plots y as a function of x
```

This series of commands produces a window (labeled "Figure No. 1") on your screen, which can be copied and pasted into a word-processing document (like this one) as follows:

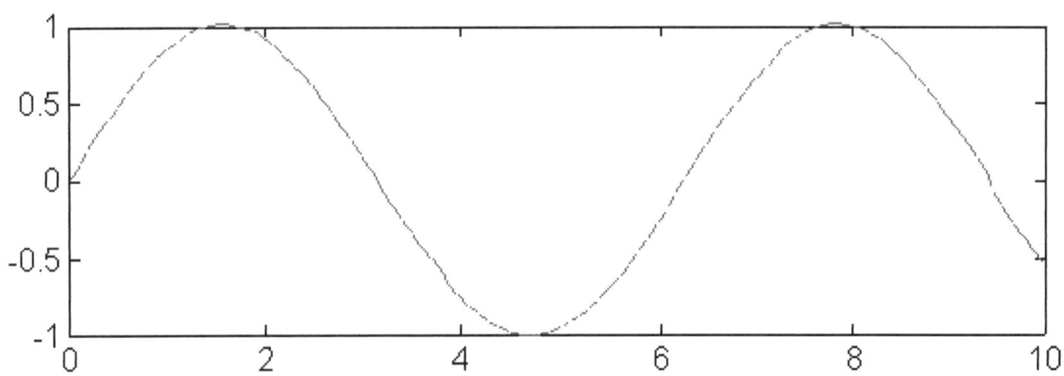

The plot function has many features, such as plotting more than one function on the same graph, using different colors and symbols, adding labels, etc., which will be presented later in the examples that follow. (Or if you are in a hurry, look it up on the MATLAB web site.)

Solving an Algebraic Equation

MATLAB has several ways to compute when a function has a value of zero. This is essentially what is needed to solve equations involving polynomials or transcendental equations which often occur in engineering. If the function is defined symbolically, the command `fzero` may be used to produce a list of zero-crossings (or roots) of the function. The second method is useful when an initial guess of the solution is available (from plotting), and in this case the function must be defined numerically. Here we examine the numerical approach to finding the solution of the equation

$10 - 10\ \sin(x/10) - x\ \cos(x/10) = 0.$

First define the equation as a function by creating the function m-file following the steps described earlier (i.e., select **New** and **M-File** from your menu). Use the name `dload`, with the first line indicating that it is a function of `x`. Note that in defining the derivative expression numerically, you need to use the appropriate multiplication symbol (`.*`) for array arithmetic. The following window illustrates the appropriate m-file to use in this case:

```
function y = dload(x)
                        % this file defines the function dload as a function of x by:
y = 10- 10*sin(x/10)-x.*cos(1/10*x);
```

Give this file the name `dload.m`, and return to the workspace window. Now, to use the command `fzero`, you must guess the value of x in the interval of interest ($0 < x < 10$, in this case). This guess is usually obtained by plotting the function and visually determining the value of x. The function is actually plotted later on, and from that plot, a guess of `x = 6` is obtained. The following code finds the exact value:

```
EDU> x0=fzero('dload',6)   % look for a zero near x =6 of the function dload
x0 =
     5.5597
```

Note that this function is periodic and has lots of zero values, so that if we typed in $x = 22$ instead of $x = 6$, we would find a different value of `x0`. Plotting the function assists in sorting out which zero we are interested in.

So far, we have introduced some basic MATLAB functions and manipulations without a lot of detail. More detail is presented "on the fly" in the pages that follow, as the various samples in the text are worked out in MATLAB. This is intended to help you work through your homework using MATLAB as an aid in making your computations so that you may focus on learning mechanics. Using MATLAB instead of a calculator will allow you to do much more and to paste your results directly into a word-processing document to give you very professional-looking homework to turn in. Combining word processing and MATLAB in this way is also very much the format in industry and will help get you ready for a summer job and the eventual engineering job after you graduate. MATLAB graphs and results can be copied and placed in Microsoft Word documents for finished reports.

Solving a Statics Problem by Using MATLAB

In this section, all of the previously defined commands are combined to solve a simple problem in statics: Given a loading function on a beam, determine where the load is a maximum. The solution comes from calculus, which certifies that the maximum or minimum value of a function occurs where its derivative vanishes. If we plot the function, we can visually determine whether it is a maximum or a minimum. If we plot the derivative, we can obtain a rough guess of where it is zero and hence where the load is a maximum. The function examined here is not too difficult and can easily be handled by hand or by using a calculator. However, using MATLAB allows you to keep a printed and digital record of your work. Furthermore, this example will serve to help you solve more complex loads and statics problems later on. The MATLAB steps used to solve the problem consist of defining the function symbolically, plotting the function, differentiating it symbolically, plotting the derivative function to get a visual guess at the point where the derivative is zero, defining the derivative function numerically, and using a root finder, along with the initial guess obtained from the plot, to calculate the value of x where the load is maximum. The maximum value of the function is determined by setting the derivative of the function equal to zero and solving for the value of x at that point. Each of these procedures has already been outlined, and all are now used together to solve a specific loading problem.

The specific problem is as follows: Given that a 10-m-long beam is loaded by the function

$$w(x) = 10\,(x - x\sin(x/10))\ \mathrm{N/m},$$

calculate the position along the beam (i.e., the specific value of x) where the load $w(x)$ has its largest value. Assume that the beam starts at $x = 0$. The first order of business is to enter the load function given for the beam. Note that in order to be able to make changes to this function later, we need to save it as an m-file. Recall that anything that has been defined in the m-file can be used in future calculations and can be easily edited. The function $w(x)$ can be plotted numerically or symbolically. Since we need to compute the derivative of $w(x)$, we will define it symbolically. (Numerical derivatives often have difficulties associated with them.) Following the directions given earlier for creating an m-file (i.e., select **New** and **M-File** from your menu), typing, and saving the following m-file defines the symbolic form of the load function, differentiates the function and plots both the function, and its derivative (choose a name, say, `loplt.m`):

```
% this sets up the beam loading function for plotting
L=10;
x=sym('x');              % this declares x as a symbolic variable
w=10*(x-x*sin(x/L));     % this is the load function to be examined
dw=diff(w);              % this is the derivative with respect to x
ezplot(w,[0,10])         % this uses the symbolic plot function to plot w(x)
                           between 0 and 10
figure                   % this instruction tells the code to leave the above plot
                                % displayed when forming the next plot
ezplot(dw,[0,10])        % this plots the derivative dw(x) between 0 and 10
```

Typing the command `lop lt` in the command window after the EDU> prompt will produce the output of this file, which yields the following two plots and the analytical expression for the derivative of $w(x)$:

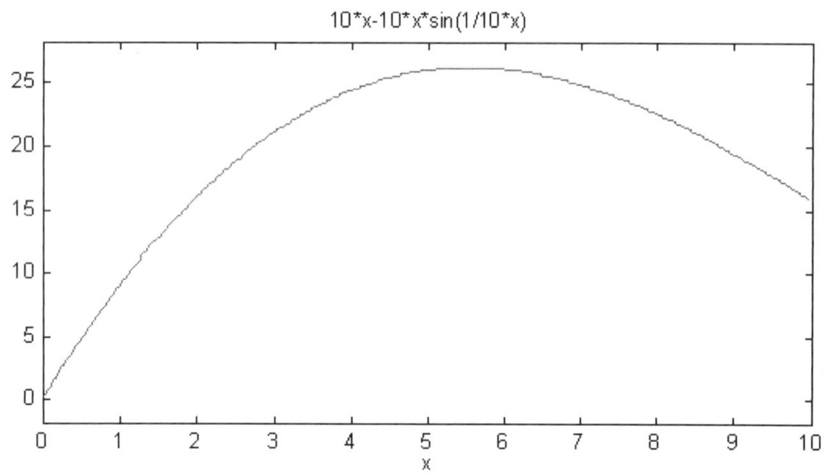

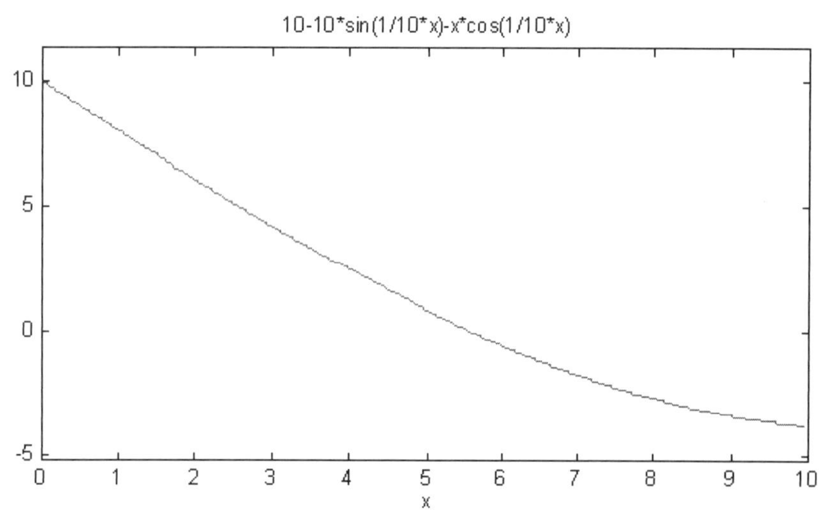

These plots have again been copied out of MATLAB and pasted into this document. Note that the `ezplot` command automatically prints the function across the top in the title area and the independent variable under the x-axis. More on this feature can be obtained by typing `help ezplot` or looking in the manual.

Examining the first plot reveals that the derivative becomes zero somewhere near the value $x = 5.8$. This provides an initial estimate of the solution of the original function. The numerical definition of the derivative is given in the m-file `dload`. Now the "`fzero`" command can be used to find the zero crossing of the `dload` function. The command requires the function name and the initial guess, as indicated in the following workspace file:

```
EDU>x0=fzero('dload', 5.8)   % look for a zero near x =5.8 of the function dload
x0 =
      5.5597
```

This code returns the value of x at which the derivative of the load function is zero, and it names that value `x0`. From calculus and the fact that the plot of the load function exhibits a peak in the region of x between 0 and 10, we know that this corresponds to the value of x where the load function is a maximum value. To find the value of the load function at its maximum, simply call `wload` in the workspace file, and evaluate it at `x0 = 5.5597` as follows:

```
EDU>wload(x0)
ans =
      26.2547
```

2

Vectors

In this chapter, we show the detailed solution of sample problems for which MATLAB is useful, demonstrate how MATLAB can be used as a vector calculator, and indicate how systems of linear equations can be solved using matrix methods. Most of the problems in the chapter do not require the use of MATLAB, but you are encouraged to solve both the vector equations and the linear systems with MATLAB, as this will give you more time to concentrate on the free-body diagrams and forming the equations of equilibrium. You will also find that numerical errors will be effectively eliminated. The most efficient use of MATLAB can be obtained by following these steps in your solution strategy: (a) draw a free-body diagram of the particle, and show all forces acting on the particle and any specified geometry; (b) express each force vector in the coordinate system chosen; (c) write the required vector equation (resultant or equilibrium); and (d) enter all this information into your MATLAB command window or m-file and use MATLAB to perform all numerical calculations.

Sample Problem 2.5

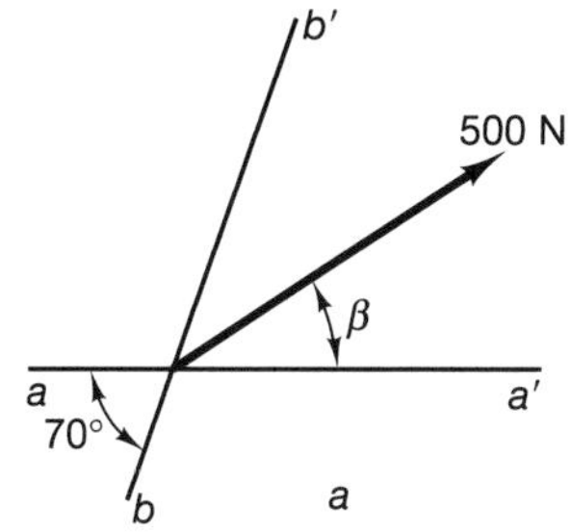

A 500-N (5.0×10^2 N) force is to be resolved into components along lines *a–a'* and *b–b'*. Determine the angle β and the component along *b–b'* if it is known that the component along *a–a'* is 320 N.

SOLUTION

Draw a force vector diagram of the force and the two components. The angle between $\boldsymbol{F}_a$ and $\boldsymbol{F}_b$ is $(180 - 70) = 110°$. Therefore, by the law of sines

$$\frac{500}{\sin(110°)} = \frac{F_b}{\sin(\beta)} \Rightarrow 500\sin\beta = F_b\sin(110°)$$

$$\frac{500}{\sin(110°)} = \frac{320}{\sin(180 - 110 - \beta)} \Rightarrow 500\sin(70 - \beta) = 320\sin(110°)$$

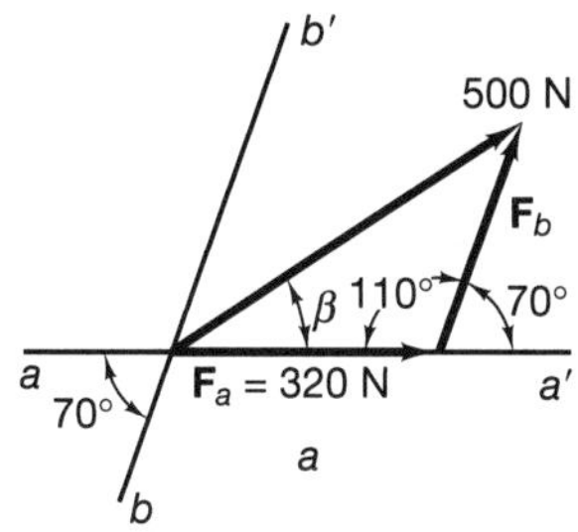

Now, $\sin(110°) = \sin(70°)$ and $\sin(70 - \beta) = \sin(70)\cos\beta - \cos(70)\sin\beta$; therefore, the second equation may be written as

$$500\sin(70)\cos\beta - 500\cos(70)\sin\beta = 320\sin(70)$$

This is a transcendental equation for β and can be solved by taking the $\sin\beta$ term to the right side and squaring both sides of the equation. The trignometric identity $\cos^2\beta = 1 - \sin^2\beta$ can then be used to form a quadratic equation in $\cos\beta$. Once the angle has been determined, the magnitude F_b is obtained from the first equation, yielding

$$\beta = 33 \text{ degrees} \qquad F_b = 290 \text{ N}$$

This sample problem requires the solution of a transcendental equation. The equation can be solved by hand, as shown in the text by using trigonometric identities to obtain a quadratic equation for $\cos\beta$, which has two roots. However, many algebraic equations of this nature cannot be easily solved and are best approached numerically. To solve such an equation numerically, first graph the function and determine the approximate roots (following the commands from Section 1). Use the first approximate root to determine the initial guess for the `fzero` function. Then iterate the `fzero` function on this initial guess until it converges on a root (also called a zero) of the function. The graph is an important first step used to specify the region in which the iteration begins. If a second root is desired, specify the initial guess in the region of the second root. Here we also introduce how to add labels and grids to plots made in MATLAB.

The solution of the transcendental equation from Sample Problem 2.5 using MATLAB is shown here using both the command window, or workspace, and m-files. To gain familiarity with graphing and the fzero function, you should reproduce these steps on your own computer. In order to be able to change some of the constants, that is, the orientation of 70 degrees, or the forces (500 N or 320 N), and examine the change in the graph and in the roots, we will save the function in a function m-file. Then you can call up the m-file with the definition of the equation in it, make the changes you want, and then save them. Afterwards, go back to the command window and run the plot again. A general solution is obtained by identifying $\theta = 70$ degrees, $V = 500$ N, and $A = 320$ N. These values can then be easily altered as variables, without editing the transcendental equation itself. Remember that while calculators allow you to enter angles in either radians or degrees, computational software, including MATLAB, will accept only radians. Hence, if you want to enter the data in degrees, convert your entry using x(in rad) =x*pi/180. Note that MATLAB recognizes pi as the value π.

First set up the function in a function m-file by selecting **New** and **M-File** from the menu. Then enter the following (recall that anything typed after % is a comment):

```
% this file defines the function f(b) numerically over the range of values from
% 0 to 90 degrees and then plots the function
b=(0:2:90)*pi/180;      % this defines the range of values of beta in degrees and
                                %changes them to radians
T=70;V=500; A=320;
F=V*sin(T*pi/180)*cos(b)-V*cos(T*pi/180)*sin(b)-A*sin(70*pi/180);
                  % this defines the force
plot(b,F)         % this plots the numerical expression F versus the values in b
grid              % this turns on the background grid on the plot
xlabel("beta in radians')   %types the phrase —beta .." under the horizontal axis
title("force in Newtons vs angle')        %types the title along the top
```

Save this code in an m-file named sp2_5.m. Then, typing the following in the workspace will produce the required plot:

```
EDU> sp2_5
```

The plot produced by this command is as follows and may now be used to guess the value of b that causes the function F to be zero—the initial guess needed for the fzero function to find the exact value:

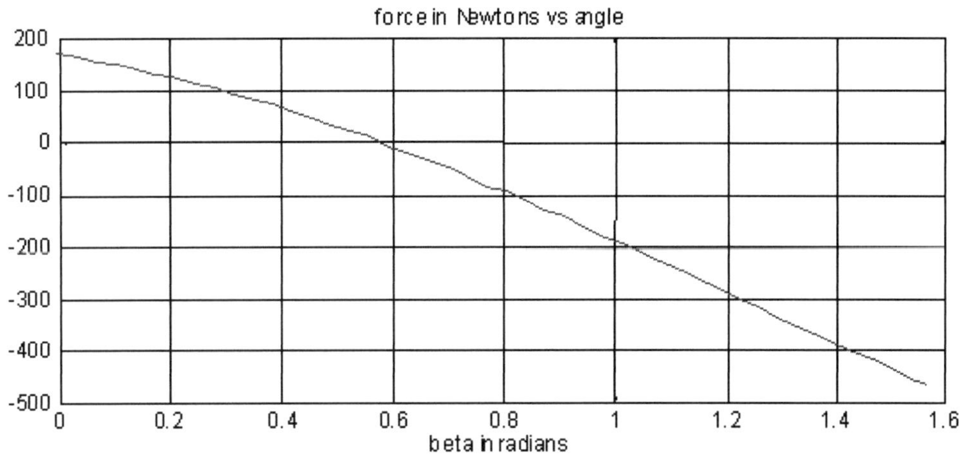

The function is plotted only in the range of physically meaningful value of β. Following the grid line at zero force until it crosses the curve provides the initial guess of b=0.6 radians as a value close to the exact solution of F = 0. Next, define F as a function in an m-file named fbeta.m as follows:

```
function F=fbeta(b)
% this sets up the fuction F of beta for use in finding a root
t=70; V=500; A=320;
F=V*sin(t*pi/180)*cos(b)-V*cos(t*pi/180)*sin(b)-A*sin(t*pi/180);
```

Now use the fzero function to find the exact value of beta by typing the following in the workspace:

```
EDU>b0=fzero('fbeta',0.6)
b0 =
      0.576
```

Using this value of b0, the force along the line B–B becomes

```
EDU>Fb=500*sin(b0)/sin(110*pi/180)
Fb=
      290.0264
```

An alternative method of ploting is to use the fplot command. This command requires only that the fuction to be plotted be previously defined in an m-file. As an example, consider plotting the function fbeta just defined. This requires only one line to be entered in the command window (unless you need to label the plot), as illustrated in the following code:

```
EDU>fplot('fbeta',[0,pi])    % this plots the function fbeta from 0 to pi
```

The resulting plot is as follows:

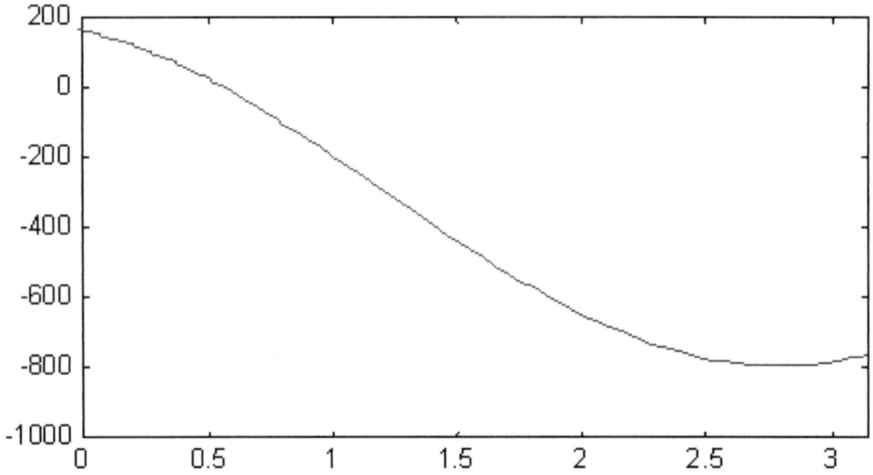

MATLAB as a Vector Calculator

Vectors in MATLAB are represented by a (3 × 1) row matrix. Numbering of the elements in a matrix starts with 1 in MATLAB. The first element in a row matrix is designated by 1, and if the vector is A, the x component is designated by A(1), which is used to call this element in MATLAB. In the text, as shown in Section 2.4, we use the notation of a column vector rather than a row vector to refer to the x,-y,-z,-components of a vector. To be compatible with the text, we can use the transpose operation to change a row vector in MATLAB into a column vector (i.e., type: A′). Alternatively, we can enter the vector as a column directly by using semicolon notation to separate the elements. Hence, the vector **A** with components $x = 2$, $y = -1.5$, and $z = 3$ is entered as follows:

```
EDU> A =[2; -1.5; 3]
A =
    2.0000
   -1.5000
    3.0000
```

The individual elements of this column vector are automatically named A(1) = 2, A(2) = -1.5, and A(3) = 3. The notation A(2) means that this is the value of the second element of the vector A, which we associate with the y-component of a vector in statics. Although the operations shown in Computational Windows 2.1 through 2.5 can be easily performed on any calculator using the scalar components, these windows serve as an introduction to the powerful tools of computational software packages. You should reproduce the contents of the window on your own until you feel comfortable making vector manipulations in MATLAB.

The specific commands for performing vector calculations are listed in this section. Vector addition is done automatically in component notation. An example of vector addition and subtraction is shown in Computational Window 2.1.

COMPUTATIONAL WINDOW 2.1: VECTOR CALCULATOR

```
EDU> [10;-5;3]+[6;5;-2]
ans=
      16
      0
      1
EDU> [10; -5; 3] - [6; 5; -2]
ans=
      4
    -10
      5
```

The elements of these vectors may contain mathematical computations:

```
EDU>B=[25*sin(pi); 25*cos(pi); sqrt(2)]
B =
           0
   -25.0000
     1.4142
```

The magnitude of the vector is computed using the command: `norm(A)` as follows:

```
EDU>norm(A)
ans =
     3.9051
```

When a vector is to be used in more than one operation, it is best to define a symbol for the vector and represent the vector operations using algebraic symbols for them. This, as well as the fact the matrix elements can contain simple mathematical calculations, is illustrated in Computational Window 2.2.

COMPUTATIONAL WINDOW 2.2: VECTOR ALGEBRA

```
EDU>A = [10; -5; 3];          B = [6; 5; -2];
EDU>A + B
ans =
    16
     0
     1
EDU>A - B
ans =
     4
   -10
     5
```

In many cases, it is useful to define the result of an operation by another symbol and to write vector equations. Note that all of the mathematics is handled numerically in these calculations:

```
EDU>C=A+B,  D=A-B,  F= 2*A-B/3
C =
    16
     0
     1
D =
     4
   -10
     5
F =
   18.0000
  -11.6667
    6.6667
```

If algebraic symbols are used to represent a vector, a unit vector along the direction of the original vector may be obtained. The magnitude of a vector can be obtained by using the `norm` command. Divide the vector **A** by its magnitude (`norm(A)`) to get the unit vector, $\hat{\mathbf{a}}$, along **A**, as illustrated in Computational Window 2.3. Although these calculations are done here using computer software, all may be done by hand and *should* be performed by hand the first few times.

COMPUTATIONAL WINDOW 2.3: CREATING A UNIT VECTOR

```
EDU>a=A/norm(A)
a =
    0.8639
   -0.4319
    0.2592
```

The vector **A** can now be written as a magnitude multiplied by its unit vector:

```
EDU>A= norm(A)*a
A =
   10.0000
   -5.0000
    3.0000
```

This, of course, is the vector **A** we started with. The components of the unit vector $\hat{\mathbf{a}}$ are the direction cosines of the vector **A**.

The previous window illustrates that a vector can be written as a unit vector times a magnitude. This magnitude is the magnitude of the vector **A** and is computed by the `norm(A)` command in MATLAB, and the unit vector consists of the direction cosines of the vector **A**. The direction cosines specify the direction in three-dimensional space that corresponds to the direction of the vector **A**. Hence, writing the vector **A** as a magnitude times a unit vector is the three-dimensional version of the often-stated definition used in two-dimensional space that a vector is a specified magnitude and direction (e.g., 100 N at 30 degrees from the x-axis). The two-dimensional idea that a vector is a magnitude and direction is often preprogrammed into calculators.

Vector algebra may also be performed symbolically. First the elements of the vector must be designated as symbolic quantities. Then these quantities can be manipulated, as Computational Window 2.4 illustrates:

COMPUTATIONAL WINDOW 2.4: SYMBOLIC PROCESSOR

```
EDU>syms Ax Ay Az Bx By Bz      % defines these quantities as symbolic
                                % (alternative to sys('x'))
EDU>A = [Ax;Ay;Az], B= [Bx;By;Bz]  % defines the symbolic vectors A and B
A =                             B =
[ Ax]                           [ Bx]
[ Ay]                           [ By]
[ Az]                           [ Bz]
```

```
EDU>C=A+B                    % symbolically add two vectors
C =
[ Ax+Bx]
[ Ay+By]
[ Az+Bz]
EDU>syms g                   % defines g as a symbolic constant
EDU>g*A
ans =
[ g*Ax]
[ g*Ay]
[ g*Az]

EDU>N=svd(A);N(3) % computes magnitude of the vector A and stores it in N(3)
ans =
     (Ax*conj(Ax)+Ay*conj(Ay)+Az*conj(Az))^(1/2)
```

Here conj refers to the complex conjugate, which can be ignored in this case because our vectors are real valued.

Other symbolic commands are available, and the *Student Version* web site should be consulted for a listing. Many other symbolic manipulations are possible. For example, the calculations just made can be used to find an analytical expression for the elements of a unit vector, which are in fact the direction cosines of the vector. This is illustrated in the following list of commands and responses.

```
EDU>N=svd(A);  % defines a vector N whose 3rd element is the magnitude of A
EDU>a=A/N(3)        % defines a unit vector along the vector A
a =
 [ Ax/(Ax*conj(Ax)+Ay*conj(Ay)+Az*conj(Az))^(1/2)]
 [ Ay/(Ax*conj(Ax)+Ay*conj(Ay)+Az*conj(Az))^(1/2)]
 [ Az/(Ax*conj(Ax)+Ay*conj(Ay)+Az*conj(Az))^(1/2)]
```

Sample Problem 2.7

Find the resultant **C** of the two vectors

$$\mathbf{A} = 6\hat{\mathbf{i}} + 2\hat{\mathbf{j}} + 0\hat{\mathbf{k}}$$
$$\mathbf{B} = 0\hat{\mathbf{i}} + 7\hat{\mathbf{j}} + 5\hat{\mathbf{k}}$$

Write vector **C** as the product of a magnitude and a unit vector in the direction of **C**.

This sample problem asks that the resultant of the addition of two vectors and the unit vector in the direction of the resultant be determined. Sample Problem 2.8 involves a greater number of vector calculations and is an excellent example of the use of computational software. The MATLAB solutions of these problems are respectively shown in the next two windows.

COMPUTATIONAL SOLUTION: SAMPLE PROBLEM 2.7

```
EDU>A = [ 6; 2; 0 ]; B = [ 0; 7; 5]; % define the vectors A and B
EDU>C = A + B                        % compute and define the sum C
C =
     6
     9
     5
EDU>norm(C)                          % compute magnitude of the vector C
ans =
   11.9164
EDU>c = C/norm(C)                    % compute the unit vector along C
c =
    0.5035
    0.7553
    0.4196
```

Sample Problem 2.8

An electronic scoreboard in a gymnasium is supported by three cables as shown in the diagram at the right. The tension in each cable is

$$T_A = 368 \text{ N}$$

$$T_B = 259 \text{ N}$$

$$T_C = 482 \text{ N}$$

Specify the force vector for each cable in terms of the magnitude of the tension and a unit vector along the cable. Then determine the resultant of the tensions.

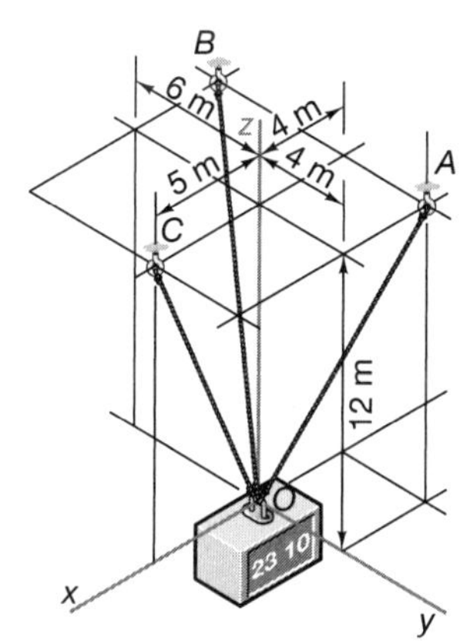

COMPUTATIONAL SOLUTION: SAMPLE PROBLEM 2.8

```
EDU>A=[-4;4;12]; B=[-4;-6;12]; C=[5;0;12];   % defines position vectors to
                                                  % the point of attachment
EDU>a=A/norm(A),b=B/norm(B),c=C/norm(C)      % computes the unit vectors
                                                  % along each position vector
a =
   -0.3015
    0.3015
    0.9045
b =
   -0.2857
   -0.4286
    0.8571
c =
    0.3846
         0
    0.9231
EDU>TA=368*a,TB=259*b,TC=482*c        % uses the unit vectors to compute the
                                      % tensions
TA =
 -110.9562
  110.9562
  332.8685
TB =
   -74
  -111
   222
TC =
  185.3846
         0
  444.9231
EDU>R=TA+TB+TC            %computes the resultant of the 3 cable forces in Newtons
R =
    0.4284
   -0.0438
  999.7916
```

Solution of Simultaneous Linear Equations

A system of linear equations is easily solved using matrix methods from linear algebra, as shown in Section 2.7 of the *Statics* text. Numerous problems in statics involve a system of linear equations, and although they can be solved by hand, many calculators and virtually all computational software packages will numerically or symbolically solve such systems of equations.

The next window shows an example of the MATLAB solution of Eq. (2.91). Many handheld calculators have the capability of handling matrices up to (6×6), i.e., the ability to solve six equations with six unknowns. MATLAB can solve systems as large as (128×128) in the *Student Edition* and larger in the professional version. Consider the following system of three linear equations with the three unknowns x, y, and z:

$$\begin{aligned} 2x + 3y - z &= 6, \\ 3x - y + z &= 8, \\ -x + y - 2z &= 0, \end{aligned}$$

In the following window the coefficient matrix is denoted by **A** and the vector on the right side is denoted by **b**.

```
EDU>A=[2 3 -1;3 -1 1;-1 1 -2]   % enters the matrix of coefficients
A =
     2     3    -1
     3    -1     1
    -1     1    -2
EDU>b=[6;8;0];                  % enters the vector of the right side
EDU>X=inv(A)*b                  % computes the solution
X =
    3.0667
   -0.6667
   -1.8667
```

Note that the first value of X is x, the second is y, and the third is z.

Sample Problem 2.9

If $\hat{\mathbf{a}}$ and $|\mathbf{B}|$ are known, determine $\hat{\mathbf{b}}$ and $\mathbf{A}$ such that $\mathbf{A} + \mathbf{B} = \mathbf{C}$.

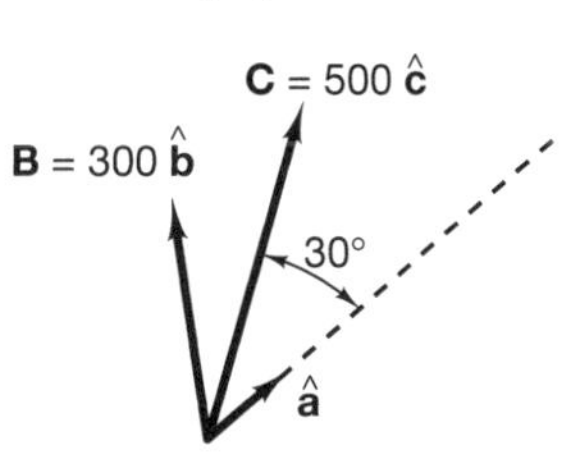

SOLUTION

From the drawing, establish a coordinate system such that $\hat{\mathbf{i}} = \hat{\mathbf{a}}$. Then the vectors may be written in this coordinate system as:

$$\mathbf{C} = 500(\cos 30°\hat{\mathbf{i}} + \sin 30°\hat{\mathbf{j}})$$

$$\mathbf{A} = A\hat{\mathbf{i}}$$

$$\mathbf{B} = 300(\lambda_x\hat{\mathbf{i}} + \lambda_y\hat{\mathbf{j}})$$

Writing the sum $\mathbf{A} + \mathbf{B} = \mathbf{C}$ in component form and equating the coefficients of $\hat{\mathbf{i}}$ and then the coefficients of $\hat{\mathbf{j}}$ yields the two equations:

$$A + 300\lambda_x = 500 \cos 30°$$

$$300\lambda_y = 500 \sin 30°$$

In addition the direction cosines must satisfy $\lambda_x^2 + \lambda_y^2 = 1$. Thus we have a system of 3 equations in 3 unknowns. Solving these three equations yields

$$\lambda_y = 0.834, \lambda_x = \sqrt{1 - \lambda_y^2} = \pm 0.553$$

$$A_+ = 267, A_- = 599$$

Note that there are two solutions.

Sample Problem 2.10

A tripod supports a 10-kg movie camera. (See figure at right.) Because this weight is transmitted down the legs of the tripod, the resultant of the three leg forces is 10 kg downward. Determine the force on each leg of the tripod by considering the legs as nonorthogonal components of the resultant vertical force.

Many problems in statics reduce to the solution of a series of simultaneous equations. Sample Problem 2.10 is another example of such a problem. In the past, problems of this nature were solved by hand by eliminating all the unknowns but one in one equation and solving this equation for the first unknown. This unknown is then substituted into the other equations. The process is repeated, and the unknowns are determined one by one. The following window illustrates how to do this using matrix methods in MATLAB.

COMPUTATIONAL SOLUTION: SAMPLE PROBLEM 2.10

```
>>A=[-60;-100;0]; B=[52.5;-100;30];C=[52.5;-100;-30]; % form vectors from
                                                        camera to base along
                                                        each leg
>>a=A/norm(A);b=B/norm(B);c=C/norm(C); % forms unit vectors along legs
>>C=[a b c]     % forms the coefficient matrix
C =
   -0.5145    0.4493    0.4493
   -0.8575   -0.8557   -0.8557
         0    0.2567   -0.2567
>>L=[0;-10*9.81;0];                % define the load vector
>>F=inv(C)*L                       % computes the solution
F =
   53.3882
   30.5706
   30.5706
```

In the window, the 1st element of **F** is the force along **A**, the second is that along **B**, and the third element is the force along **C**.

Solving systems of linear equations by using the `X=inv(A)*b` command is not numerically the most efficient method, because computing the inverse of a matrix is sometimes ill conditioned, even when a perfectly fine solution exists. Instead, MATLAB offers an important alternative that should be used. The backslash, found three keys to the left of the P key, computes the solution to a set of linear equations using LU factorization, which is a numerically superior method to using `X=inv(A)*b`. (See *The Student Version* or the MATLAB web site for more details.) To implement LU factorization, use the command `X=A\b` instead of typing `X=inv(A)*b`. Besides giving you a numerically superior result, this combination of characters is faster to type! Try repeating the previous computations with `X=A\b` instead.

Using MATLAB in Other Matrix Calculations

Although Sample Problem 2.12 can be solved by hand, all numerical operations can be performed using MATLAB. Most legitimate linear algebra calculations also can be carried out using MATLAB, as illustred in Sample Problem 2.12.

Sample Problem 2.12: A MATLAB Solution

For the two matrices

$$[A] = \begin{bmatrix} 2 & -3 & 1 \\ 0 & 2 & -1 \\ 3 & 1 & 1 \end{bmatrix} \quad [B] = \begin{bmatrix} 5 & 3 & 2 \\ -2 & 1 & 4 \\ -1 & 0 & -1 \end{bmatrix}$$

determine: (a) $[A] + [B]$
(b) $[A][B]$
(c) $[B][A]$

The following list of commands forms the MATLAB solution for Sample Problem 2.12 (note that the matrices are first entered and stored as symbols (**A** and **B**) and then manipulated):

```
EDU>A=[2 -3 1;0 2 -1;3 1 1]; % enter A a row at a time, separated by semicolons
EDU>B=[5 3 2;-2 1 4;-1 0 -1]; % enter B
EDU>A+B
ans =
     7     0     3
    -2     3     3
     2     1     0
EDU>A*B
ans =
    15     3    -9
    -3     2     9
    12    10     9
EDU>B*A
ans =
    16    -7     4
     8    12     1
    -5     2    -2
```

Scalar or Dot Product

The scalar, or dot, product between two vectors A and B can be obtained by use of the command dot(A,B) or by taking the transpose of one vector and using the * command, i.e., A'*B. This operation may be performed numerically or symbolically in MATLAB. Both symbolic and numerical calculations are given in the next window. The dot product can be used to compute the angle between two vectors and can also be used to find components and projections, as discussed in the *Statics* text. The command windows that follow illustrate the use of the dot product. First, consider the dot product of the two vectors **A** and **B** using numerical evaluation:

```
EDU>A = [3; 7; -1]; B=[-5; 4; 9];    % enters the vectors of interest
EDU>dot(A,B)                          % computes the dot product numerically
ans =
     4
EDU>A'*B                              % alternative calculation of the dot product
ans =
     4
```

Next, consider computing the dot product symbolically (note that the conjfunction returned by MATLAB refers to the complex conjugate, which is the same as the real part here because the vectors are all real valued):

```
EDU>syms Ax Ay Az Bx By Bz            % defines these quantities as symbolic
EDU>A=[Ax;Ay;Az];B=[Bx;By;Bz];        % defines the vectors A and B symbolically
EDU>dot(A,B)
ans =
conj(Ax)*Bx+conj(Ay)*By+conj(Az)*Bz
```

The numerical values of **A** and **B** and the dot product can be used to calculate the angle between **A** and **B**:

```
EDU> angle = acos(dot(A,B)/(norm(A)*norm(B)))    % use the arc cosine command
angle =
    1.5236
EDU>1.5236*180/pi                                 % changes radians to degrees
angle =
   87.2958
```

Sample Problems 2.13

These problems are solved using MATLAB, as each requires the use of specific vector operations only.

Two vectors, shown in the diagram at the right, are given as:

$$\mathbf{A} = 3\hat{\mathbf{i}} - 2\hat{\mathbf{j}} - 2\hat{\mathbf{k}}$$

$$\mathbf{B} = 4\hat{\mathbf{i}} + 4\hat{\mathbf{j}}$$

Sketch the two vectors in space and determine the angle between them.

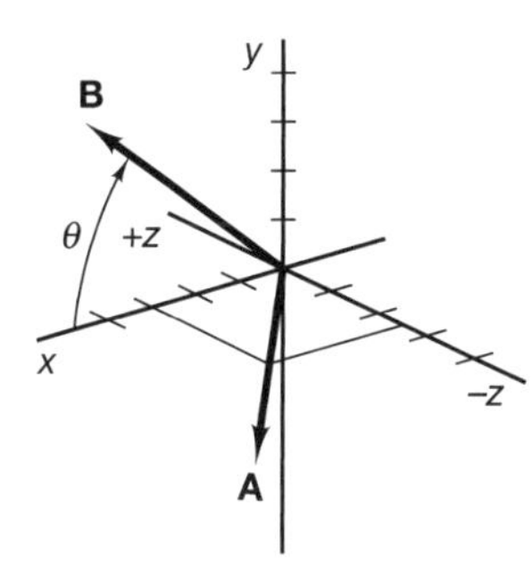

COMPUTATIONAL SOLUTION: SAMPLE PROBLEM 2.13

```
EDU>A=[3 ;-2; -2]; B= [4; 4; 0];                % enters the two vectors of interest
EDU>angle=acos(dot(A,B)/(norm(A)*norm(B)))      % compute the angle
angle =
    1.3984
EDU>angdeg=angle*180/pi                         % change the angle to degrees
angdeg =
   80.1250
```

Sample Problems 2.14

A force $\mathbf{F} = 10\hat{\mathbf{i}} - 10\hat{\mathbf{j}} + 5\hat{\mathbf{k}}$ N is applied to a bar, whose position is shown in the diagram at the right. Determine the component of force that is transmitted along the bar and the component that is perpendicular to the bar.

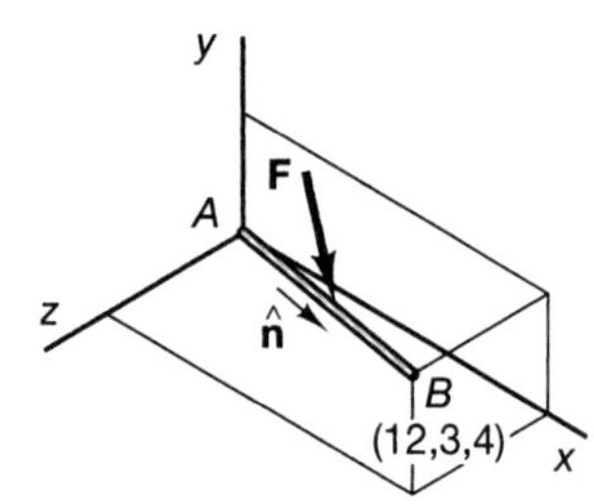

COMPUTATIONAL SOLUTION: SAMPLE PROBLEM 2.14

```
EDU>F=[10; -10; 5]; Rba = [12; 3; 4];    % enters the force and position vectors
EDU>n =Rba/norm(Rba);                    % computes the unit vector along Rba
EDU>Fab=dot(F,n)*n                       % computes the vector component along
                                         % the line
Fab =
    7.8107
    1.9527
    2.6036
EDU>Fperp= F-Fab                         % Computes the component perpendicular
                                         % to the line
Fperp =
    2.1893
  -11.9527
    2.3964
```

Vector or Cross Product

MATLAB calculates the cross product between two vectors A and B by using the command `cross(A,B)`. It is important to know how to perform cross products by hand when the vectors are fairly simple. But when the vectors are more complicated, using a machine is likely to produce fewer numerical errors, as well as give quicker results. This is demonstrated in Computational Windows 2.5 and 2.6.

COMPUTATIONAL WINDOW 2.5

The dot product can be obtained by a matrix multiplication between two vectors written as column or row matrices. By contrast, the cross product cannot be obtained through any standard form of matrix multiplication. Consider three vectors **A**, **B**, and **C** written as column matrices:

```
EDU>A=[3;-1;2]; B=[-1;5;2]; C=[2;1;2]; % enters 3 column vectors of interest
EDU>cross(A,B)     % computes the vector A cross B
ans =
   -12
    -8
    14
EDU>cross(B,A)     % computes the vector B cross A
ans =
    12
     8
   -14
EDU>dot(A,cross(B,C))        % computes  the number A dot (B cross C)
ans =
    -4
EDU>dot(B,cross(C,A))        % computes the number B dot (C cross A)
ans =
    -4
EDU>cross(A,cross(B,C))      % computes the vector A cross (B cross C)
ans =
    -1
    49
    26
EDU>B*dot(C,A)-C*dot(A,B)    % computes the vector B times (C dot A) - C times
                             % (A dot B)
ans =
    -1
    49
    26
```

These vector products can be placed in an m-file, rather than performed directly in the command window. The m-file allows you to edit the computational steps, in case you enter the wrong number into one element of a vector and you need to change it. Also, if you are looking at "what if" questions and need to change an element of a vector an m-file is a better choice. Next, consider using the symbolic approach to performing the vector cross and dot products, as described in Computational Window 2.6.

COMPUTATIONAL WINDOW 2.6

The vector dot and cross products are valid operations in the MATLAB symbolic code and are useful in looking at the various products that occur in mechanics.

```
EDU>syms Ax Ay Az Bx By Bz Cx Cy Cz   % define the vector entries as symbolic
                                      % variables
EDU>A=[Ax;Ay;Az];B=[Bx;By;Bz];C=[Cx;Cy;Cz];   % define the column vectors A, B,
                                              % and C
EDU>cross(A,B)      % compute the cross product of A and B (just as in the
                    % nonsymbolic case)
ans =
[ Ay*Bz-Az*By]
[ Az*Bx-Ax*Bz]
[ Ax*By-Ay*Bx]
EDU>dot(A,cross(B,C))   % compute A dot B cross C, called the triple scalar
                        % product
ans =
conj(Ax)*(By*Cz-Bz*Cy)+conj(Ay)*(Bz*Cx-Bx*Cz)+conj(Az)*(Bx*Cy-By*Cx)
```

% Note that, as before, conj(Ax) stands for the complex conjugate of
% Ax, and since our values are all real, this is just the number Ax.

```
EDU>cross(A,cross(B,C))      % compute the vector triple product
ans =
[ Ay*(Bx*Cy-By*Cx)-Az*(Bz*Cx-Bx*Cz)]
[ Az*(By*Cz-Bz*Cy)-Ax*(Bx*Cy-By*Cx)]
[ Ax*(Bz*Cx-Bx*Cz)-Ay*(By*Cz-Bz*Cy)]
```

The answers appear as elements of column vectors, listed in descending order.

Again, this series of commands can also be written in an m-file to keep a permanent record of your program and to allow convenient editing.

Sample Problem 2.16

Consider two vectors

$$\mathbf{A} = 5\hat{\mathbf{i}} + 3\hat{\mathbf{j}}$$

and

$$\mathbf{B} = 3\hat{\mathbf{i}} + 6\hat{\mathbf{j}}$$

Determine: (a) $\mathbf{A} + \mathbf{B}$
(b) $\mathbf{A} \cdot \mathbf{B}$
(c) the angle θ between A and B
(d) $\mathbf{A} \times \mathbf{B}$ (compare the magnitude of the cross product to the magnitude given in the definition $|\mathbf{A}||\mathbf{B}| \sin \theta$)
(e) $\mathbf{B} \times \mathbf{A}$

This problem requires that vectors be added and multiplied as illustrated in the following code.

COMPUTATIONAL SOLUTION: SAMPLE PROBLEM 2.16

%Create the vectors **A** and **B** as column matrices with the *z*-components entered as zero:

```
EDU>A=[5;3;0]; B=[3;6;0];
EDU>A+B
ans =
     8
     9
     0
EDU>dot(A,B)
ans =
    33
EDU>norm(A)
ans =
    5.8310
EDU>norm(B)
ans =
    6.7082
EDU> (acos(dot(A,B)/(norm(A)*norm(B))))*180/pi
ans =
   32.4712
```

```
EDU>cross(A,B)
ans =
      0
      0
     21
EDU>cross(B,A)
ans =
      0
      0
    -21
```

Sample Problem 2.17

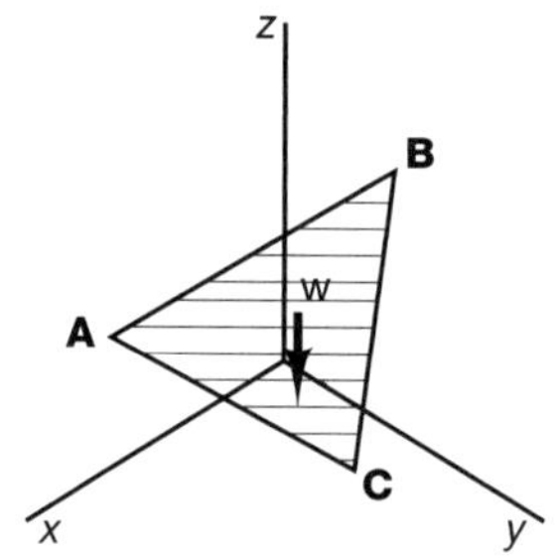

Determine the components of the 50-kg mass as a weight vector perpendicular and tangent to the plane formed by points $A(4,0,2)$ m, $B(0,2,4)$ m and $C(1,3,0)$ m. The x-y plane is horizontal and the z axis is vertical.

SOLUTION

Any three non-collinear points in space form a plane. First sketch the 3-dimensional axes and locate the points A, B, and C. Sketch the plane formed by these three points to obtain a conceptual feel for the problem.

First form vectors from point B to point A and from point B to point C:

$$\mathbf{r}_{A/B} = \mathbf{r}_A - \mathbf{r}_B = 4\hat{\mathbf{i}} - 2\hat{\mathbf{j}} - 2\hat{\mathbf{k}}$$
$$\mathbf{r}_{C/B} = \mathbf{r}_C - \mathbf{r}_B = \hat{\mathbf{i}} + \hat{\mathbf{j}} - 4\hat{\mathbf{k}}$$

By definition, the cross product of these two vectors is normal to the plane that they lie in. Thus a unit vector normal to the plane is:

$$\hat{\mathbf{n}} = \frac{\mathbf{r}_{A/B} \times \mathbf{r}_{C/B}}{|\mathbf{r}_{A/B} \times \mathbf{r}_{C/B}|} = 0.549\hat{\mathbf{i}} + 0.768\hat{\mathbf{j}} + 0.329\hat{\mathbf{k}}$$

The weight vector $\mathbf{W}$ is: $\mathbf{W} = -50 \cdot 9.81\hat{\mathbf{k}}$ N. The component of $\mathbf{W}$ normal to the plane is obtained by using the dot product with the unit normal:

$$\mathbf{W}_n = (\mathbf{W} \cdot \hat{\mathbf{n}})\hat{\mathbf{n}} = -88.6446\hat{\mathbf{i}} - 124.1024\hat{\mathbf{j}} - 53.1867\hat{\mathbf{k}}$$

The tangential component is the remaining part of $\mathbf{W}$ obtained by subtracting off the normal part:

$$\mathbf{W}_t = \mathbf{W} - \mathbf{W}_n = 88.6446\hat{\mathbf{i}} + 124.1024\hat{\mathbf{j}} - 437.3133\hat{\mathbf{k}}$$

The vector operations described in this example are very important when considering the mechanics of objects on inclined planes supported by friction.

COMPUTATIONAL SOLUTION: SAMPLE PROBLEM 2.17

```
EDU>rA=[4;0;2];rB=[0;2;4];rC=[1;3;0];
EDU>rAB=rA-rB,rCB=rC-rB
rAB =
     4
    -2
    -2
rCB =
     1
     1
    -4
EDU>n=cross(rAB,rCB)/norm(cross(rAB,rCB))
n =
    0.5488
    0.7683
    0.3293
EDU>W=-50*9.81*[0;0;1];
EDU>Wn=(W'*n)*n
Wn =
  -88.6446
 -124.1024
  -53.1867
EDU>Wt=W-Wn
Wt =
   88.6446
  124.1024
 -437.3133
```

Sample Problem 2.18

Solve the following vector equation for the magnitudes A, B and C: $A\hat{\mathbf{a}} + B\hat{\mathbf{b}} + C\hat{\mathbf{c}} = \mathbf{D}$, where:

$$\hat{\mathbf{a}} = -0.231\hat{\mathbf{i}} - 0.308\hat{\mathbf{j}} + 0.923\hat{\mathbf{k}}, \quad \hat{\mathbf{c}} = 0.385\hat{\mathbf{i}} + 0.923\hat{\mathbf{k}}$$

$$\hat{\mathbf{b}} = -0.231\hat{\mathbf{i}} - 0.308\hat{\mathbf{j}} + 0.923\hat{\mathbf{k}}, \qquad \mathbf{D} = 100\hat{\mathbf{k}}$$

SOLUTION
Using the given unit vectors compute the three biorthogonal unit vectors as follows:

$$\mathbf{R} = \hat{\mathbf{b}} \times \hat{\mathbf{c}} = 0.231\hat{\mathbf{i}} + 0.586\hat{\mathbf{j}} - 0.118\hat{\mathbf{k}}, \quad |\mathbf{R}| = 0.646$$

$$\hat{\mathbf{r}} = \frac{\mathbf{R}}{|\mathbf{R}|} = 0.440\hat{\mathbf{i}} + 0.879\hat{\mathbf{j}} - 0.183\hat{\mathbf{k}}$$

$$\mathbf{S} = \hat{\mathbf{c}} \times \hat{\mathbf{a}} = 0.284\hat{\mathbf{i}} - 0.879\hat{\mathbf{j}} - 0.118\hat{\mathbf{k}}, \quad |\mathbf{S}| = 0.646$$

$$\hat{\mathbf{s}} = \frac{\mathbf{S}}{|\mathbf{S}|} = 0.440\hat{\mathbf{i}} - 0.879\hat{\mathbf{j}} - 0.183\hat{\mathbf{k}}$$

$$\mathbf{T} = \hat{\mathbf{a}} \times \hat{\mathbf{b}} = -0.568\hat{\mathbf{i}} - 0.142\hat{\mathbf{k}}, \quad |\mathbf{T}| = 0.568$$

$$\hat{\mathbf{t}} = \frac{\mathbf{T}}{|\mathbf{T}|} = -0.970\hat{\mathbf{i}} - 0.243\hat{\mathbf{k}}$$

With the biorthogonal vectors computed, the vector magnitudes may be obtained as follows:

$$A = \frac{\hat{\mathbf{r}} \cdot \mathbf{D}}{\hat{\mathbf{r}} \cdot \hat{\mathbf{a}}} = 33.854, \quad B = \frac{\hat{\mathbf{s}} \cdot \mathbf{D}}{\hat{\mathbf{s}} \cdot \hat{\mathbf{b}}} = 33.854, \text{ and } C = \frac{\hat{\mathbf{t}} \cdot \mathbf{D}}{\hat{\mathbf{t}} \cdot \hat{\mathbf{c}}} = 40.625$$

These calculation are time consuming but all computational software programs will perform them quickly and easily as illustrated in the following example.

COMPUTATIONAL SOLUTION: SAMPLE PROBLEM 2.18

Save the following in an m file named SP218.m

```
a=[-0.231;-0.308;0.923];b=[-0.231;0.308;0.923];c=[0.385;0;0.923];D=[0;0;100];
r=cross(b,c)/norm(cross(b,c));s=cross(c,a)/norm(cross(c,a));
t=cross(a,b)/norm(cross(a,b));
A=r'*D/(r'*a),B=s'*D/(s'*b),C=t'*D/(t'*c)
```

Then type the name of the file in the command window to get the answer:

```
EDU>SP218
A =
   33.8570
B =
   33.8570
C =
   40.6284
```

This example shows how to work in the m-file, which can be saved, rather then working in the command window where nothing can be saved. This way if an error is made in entering a number, it is easier to go back an change the entry and re save the m-file.

An Additional Sample Problem

Solve Sample Problem 2.10 using a direct vector solution.

SOLUTION

The problem is to find F_A, F_B and F_C given: $F_A\hat{\mathbf{a}} + F_B\hat{\mathbf{b}} + F_C\hat{\mathbf{c}} = -50\hat{\mathbf{j}}$ where

$$\hat{\mathbf{a}} = -0.510\hat{\mathbf{i}} - 0.857\hat{\mathbf{j}}, \hat{\mathbf{b}} = 0.449\hat{\mathbf{i}} - 0.856\hat{\mathbf{j}} + 0.257\hat{\mathbf{k}}, \hat{\mathbf{c}} = 0.499\hat{\mathbf{i}} - 0.856\hat{\mathbf{j}} - 0.257\hat{\mathbf{k}}$$

From these given unit vectors compute the three biorthogonal vectors;

$$\hat{\mathbf{r}} = \frac{\hat{\mathbf{b}} \times \hat{\mathbf{c}}}{|\hat{\mathbf{b}} \times \hat{\mathbf{c}}|} = 0.886\hat{\mathbf{i}} + 0.465\hat{\mathbf{j}}$$

$$\hat{\mathbf{s}} = \frac{\hat{\mathbf{c}} \times \hat{\mathbf{a}}}{|\hat{\mathbf{c}} \times \hat{\mathbf{a}}|} = -0.256\hat{\mathbf{i}} + 0.152\hat{\mathbf{j}} - 0.995\hat{\mathbf{k}}$$

$$\hat{\mathbf{t}} = \frac{\hat{\mathbf{a}} \times \hat{\mathbf{b}}}{|\hat{\mathbf{a}} \times \hat{\mathbf{b}}|} = -0.256\hat{\mathbf{i}} + 0.152\hat{\mathbf{j}} + 0.995\hat{\mathbf{k}}$$

The three desired magnitudes are then computed to be:

$$F_A = \frac{\hat{\mathbf{r}}\cdot(-50\hat{\mathbf{j}})}{\hat{\mathbf{a}}\cdot\hat{\mathbf{r}}} = 27.3, F_B = \frac{\hat{\mathbf{s}}\cdot(-50\hat{\mathbf{j}})}{\hat{\mathbf{b}}\cdot\hat{\mathbf{s}}} = 15.5, F_C = \frac{\hat{\mathbf{t}}\cdot(-50\hat{\mathbf{j}})}{\hat{\mathbf{b}}\cdot\hat{\mathbf{t}}} = 15.5$$

COMPUTATIONAL SOLUTION: An Additional Sample Problem

First create and save the following m-file under the name SP219.m by copying the file SP218.m and changing the data for a, b, c and D to fit this problem.

```
a=[-0.51;-0.857;0];b=[0.449;-0.856;0.257];c=[0.449;-0.856;-0.257];
D=[0;-50;0];
r=cross(b,c)/norm(cross(b,c));s=cross(c,a)/norm(cross(c,a));
t=cross(a,b)/norm(cross(a,b));
FA=r'*D/(r'*a),FB=s'*D/(s'*b),FC=t'*D/(t'*c)
```

Then type SP219 in the command window after the prompt to fetch the answer:

```
EDU>SP219
FA =
   27.3329
FB =
   15.5232
FC =
   15.5232
```

Note that by using an m-file instead of the command window, this second problem of the same type becomes very easy to solve but just entering the data for the second problem into the code written for the first problem.

3

Particle Equilibrium

Although specific dimensions and physical constants are given in most undergraduate mechanics problems, greater insight into statics problems can be obtained by seeking a general or parametric solution of the problem. An example of this is shown in the solution of the next sample problem.

Sample Problem 3.4

Consider the pulley system used to hold a 200-N bale of hay in the figure at right. Determine the force **F** and the tension **T** in the system if the angle θ is specified.

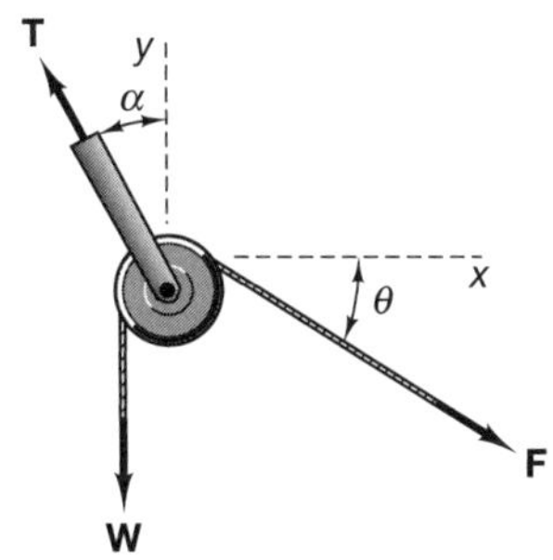

This example is also used to introduce a number of additional and useful MATLAB commands.

COMPUTATIONAL SOLUTION: SAMPLE PROBLEM 3.4

```
EDU>t=(0:5:90)*pi/180;      % sets of the range of values of theta (t) in 5
                                % degree increments, then changes this increments
                                % into radians for use in computing
EDU>n=length(t);            % determines the number of values assigned to t, for
                                % later use
EDU>F=200*[cos(t);sin(t); zeros(size(t))]; % defines a "vector' of three
                                           % components, each of which is a
                                           % function of t. Thus, this is a
                                           % matrix that is 3 x n, and each
                                           % column is the function
        % evaluated at the next time step. Note. The 3rd row establishes a 0 in
        % each element.
EDU>W=[0; 200; 0];          % defines the 3x1 loading vector
EDU>T=-W(:,ones(1,n))-F;    % the W(:,ones(1,n)) command generates the 3xn
                            % vector, each column of which looks like W of
                            % the previous line (see notes below)
EDU>magT=sqrt(sum(T.^2));   % defines the magnitude of the vector T as a function
                            % of t
EDU>magT(1)                 % this is the value of the resultant magnitude at
                            % theta = 0 deg
ans =
   282.8427
EDU>magT(5)                % this is the value of the resultant magnitude at
                           % theta = 5*(5-1)= 20 deg
ans =
   327.6608
EDU>plot(t*180/pi,magT),xlabel('theta in degrees'),
ylabel('magnitude in newtons')
```

The preceeding series of commands produces the following plot

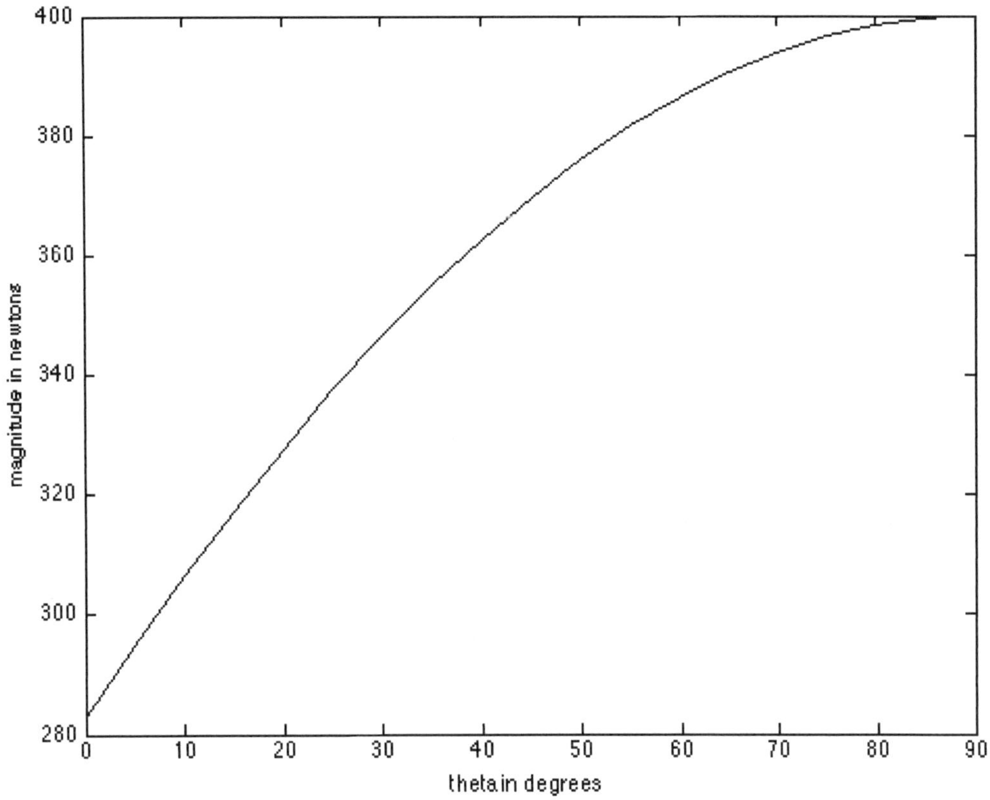

The angle α between the cable to the pulley and the vertical can also be found and plotted for any θ as is illustrated next (here lowercase a is used to denote α):

```
EDU>a=(atan(T(1,:)./T(2,:)))*180/pi;  %divides row 1 of T by row 2 element by
  % element, takes the arctangent, and converts the result to degrees
EDU>plot(t*180/pi,a)        % converts t to degrees and plots alpha as a function
  % of theta
EDU>xlabel('theta in degrees')
EDU>ylabel('alpha in degrees')
```

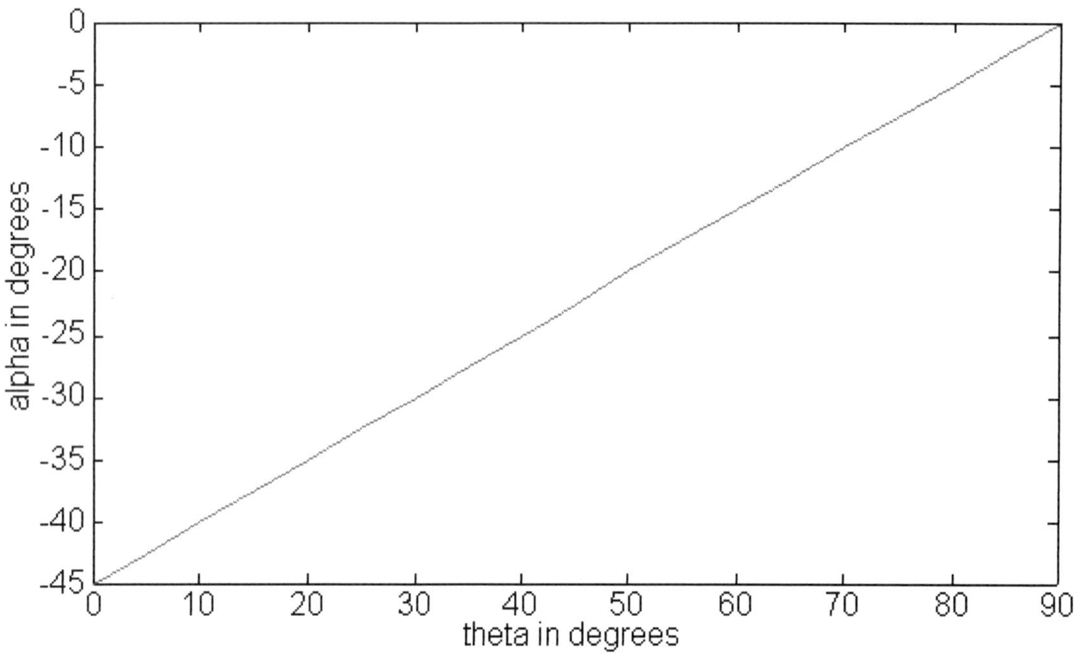

In the previous sections, we needed to introduce the idea that MATLAB stores numerical functions as vectors, as well as introduce the commands that create matrices. Rather then solving this sample problem for one value of the angle θ, we solved it for all values by naming θ as a range of values. In the problem, we use the symbol `t` rather than θ to denote the angle. The command defining `t` places the various numerical values of `t` into a $1 \times n$ vector, where n is the number of different values of the angle. The problem also introduces the use of the colon for referencing the columns or the rows of a matrix. The notation `A(:, 2)` denotes the second column of the matrix `A`. The notation `T(1,:)`, on the other hand, refers to the first row of the matrix `T`. The notation `a(j)` denotes the jth element of the row vector `a`. (See the MATLAB Web site for more on referring to elements inside an array (e.g., `T(1,:)`).) Note that each successive element of the vector `magT` corresponds to a successive value of θ. That is, the jth element of `magT` corresponds to the jth element of the vector `t`, which is `t = j*(j-1)` degrees in this case. Several other commands are used as well. The `length(t)` command produces a single number equal to the number of elements in t. The `size(t)` command yields a vector of two elements, the first the number of rows and the second the number of columns. So in the preceeding window, `length(t) = 19` and `size(t) = 1 19`. The command `zeros(2,3)` produces a matrix of zeros of dimension `2x3`, and `ones(size(X))` generates a matrix of ones the size of the matrix `X`. In working with vectors of functions, as in Sample Problem 3.4, it is very important to realize that the discrete form of a function is stored in a $1 \times n$ array, where n is the number of points at which the function is evaluated. Dealing with the matrix form of a function is sometimes tricky. The command just used to fill out the constant vector W (i.e., `W(:,ones(1,n))`) is called "Tony's Trick" and can be found (along with other hints on using the code) at the MATLAB Web site, < http://www.matworks.com/support/ tech-notes/ 1100/1109.html >.

Sample Problem 3.5

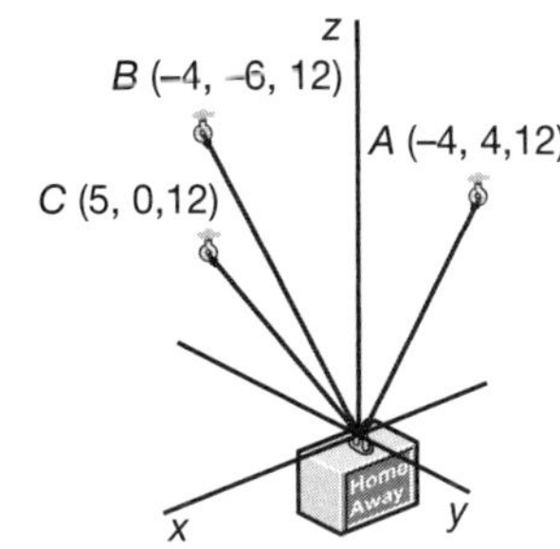

Consider a scoreboard of mass 200 kg supported by three cables, as shown at the right. Determine the tension in each of the cables.

The vector capabilities of MATLAB are used in solving Sample Problem 3.5. The three simultaneous equations are solved using matrix notation.

COMPUTATIONAL SOLUTION: SAMPLE PROBLEM 3.5

```
EDU>a=[-.302; .302;.905]; b=[-.286; -.429; .857]; c=[.385; 0; .923]; % enter the
  % 3 unit vectors
EDU>C=[a b c]     % create the matrix C by assigning its columns to be the unit
                  % vectors
C =
   -0.3020   -0.2860    0.3850
    0.3020   -0.4290         0
    0.9050    0.8570    0.9230
EDU>W=[0;0;-200*9.81]; % create the column vector corresponding to the load
EDU>T=inv(C)*(-W);     % solve the equation C T = -W, for the vector T
EDU>T(1)               % display the 1st element of T, which is the tension in A.
ans =
  722.6549
EDU>T(2)               % display the 2nd element of T, which is the tension in B.
ans =
  508.7221
EDU>T(3)               % display the 3rd element of T, which is the tension in C.
ans =
  944.7697
```

The following code solves this same problem using the direct vector approach. The vector equilibrium equation is $\mathbf{T}_A + \mathbf{T}_B + \mathbf{T}_C = \mathbf{W}$. The method requires simply calculating the unit vectors along the cables, then the biorthogonal vectors and finally the solution all by vector products with no matrix inversion.

DIRECT VECTOR SOLUTION: SAMPLE PROBLEM 3.5

Create the m-file named SP35DV:

```
A=[-4;4;12];B=[-4;-6;12];C=[5;0;12];W=[0;0;200*9.81];
a=A/norm(A);b=B/norm(B);c=C/norm(C);
r=cross(b,c);s=cross(c,a);t=cross(a,b);
Ta=W'*r/(a'*r),Tb=W'*s/(b'*s),Tc=W'*t/(c'*t)
```

Type the file name in the command window:

```
EDU>SP35DV
Ta =
  723.0242
Tb =
  508.6667
Tc =
  944.6667
```

Sample Problem 3.6

The rock climber is shifting her weight from her left leg to her right leg as she climbs. Model her as a particle with all forces concurrent with her center of mass at her hip and determine the force on her left leg and on her right leg when she is weight bearing on each leg alone.

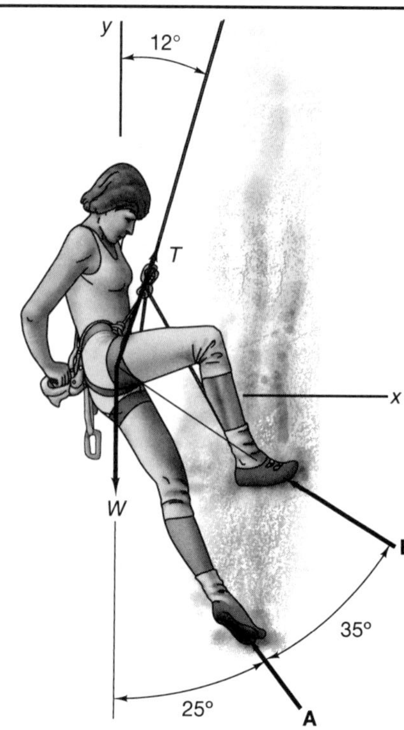

SOLUTION

The angle between the forces can be approximated on the free-body diagram as shown. The problem can be solved in terms of the weight of the climber. The vectors are expressed in components in the coordinate system chosen as horizontal and vertical in a two-dimensional plane.

$$\mathbf{T} = T(\sin 12^\circ \hat{\mathbf{i}} + \cos 12^\circ \hat{\mathbf{j}})$$

$$\mathbf{W} = -W\hat{\mathbf{j}}$$

$$\mathbf{A} = A(-\sin 25^\circ \hat{\mathbf{i}} + \cos 25^\circ \hat{\mathbf{j}})$$

$$\mathbf{B} = B(-\sin 60^\circ \hat{\mathbf{i}} + \cos 60^\circ \hat{\mathbf{j}})$$

When the left leg is the only one weight bearing, the vector equilibrium equation is:

$$\mathbf{A} + \mathbf{T} + \mathbf{W} = 0$$

This equation can be solved either by expanding into components or by the direct vector method, yielding:

$$A = 0.345\ \mathrm{W}$$

$$T = 0.703\ \mathrm{W}$$

When the weight is shifted totally to the right leg, the vector equilibrium equation is:

$$\mathbf{B} + \mathbf{T} + \mathbf{W} = 0$$

Solving the vector equation yields:

$$B = 0.219\ \mathrm{W}$$

$$T = 0.911\ \mathrm{W}$$

COMPUTATIONAL SOLUTION: SAMPLE PROBLEM 3.6

Save the following commands in an m-file named SP36.m:

```
a=[-sin(25*pi/180);cos(25*pi/180);0];b=[-sin(60*pi/180);cos(60*pi/180);0];
w=[0;1;0];t=[sin(12*pi/180);cos(12*pi/180);0];k=[0;0;1];
r=cross(b,k);s=cross(a,k);tt=cross(k,t);
A=(tt'*w/(tt'*a)),Ta=(s'*w/(s'*t)),B=(tt'*w/(tt'*b)),Tb=(r'*w/(r'*t)),
```

Then type SP36 in the command window after the prompt to get:

```
EDU>SP36
A =
    0.3455
Ta =
    0.7022
B =
    0.2186
Tb =
    0.9106
```

Solution of Nonlinear Algebraic Equations

MATLAB solves a system of nonlinear equations using the symbolic manipulation command for a small number of equations and an iterative technique (the Newton-Raphson method) for a large number of equations. The command for numerically solving a system of nonlinear equations (`fsolve`) resides in the Optimization Tool Box and is not available in the *Student Edition of MATLAB*. However, the `fsolve` function may be available if you are working under a university site license, and it should be used if available. If it not available to you at your institution, see the short version of function in the appendix under the name `mnewt.m`. Both of the commands `mnewt` and `fsolve` invoke Newton–Raphson methods of numerically solving systems of nonlinear algebraic equations. Such systems of equations have multiple solutions and require an initial guess nearest the solution of interest. (Note that using the symbolic approach does not require this initial guess.) Basically, the numerical procedure starts with an initial guess of the unknown variables and then continues to iterate until the solution converges. As might be expected, the success of the iterative technique is dependent upon the accuracy of the initial guess. The equations to be solved must all be placed in a form such that zero appears on one side of the equality. First we will look at the numerical approach. After a few sample problems are solved numerically, we examine the symbolic approach.

Sample Problem 3.8

The cable AB keeps the 10-kg collar at A in equilibrium along the smooth vertical rod as shown in the diagram at the right. Determine the tension in the cable.

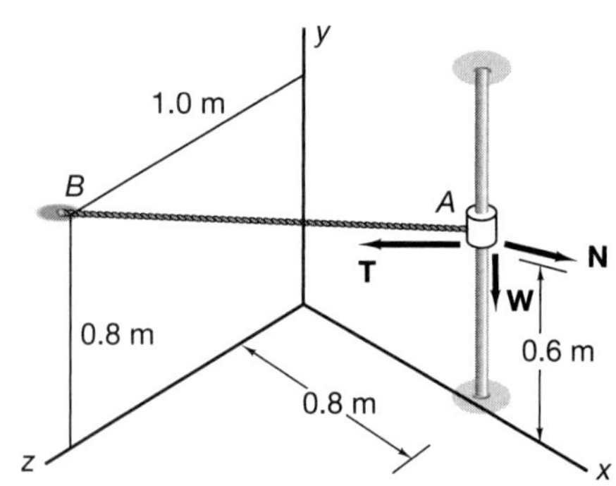

SOLUTION

Three forces act on the collar: the tension in the cable, a force normal to the rod, and the weight of the collar. These three vectors can be written as:

$$\mathbf{T} = T\hat{\mathbf{t}}$$

$$\mathbf{N} = N\hat{\mathbf{n}}$$

$$\mathbf{W} = -10*9.81\hat{\mathbf{j}}$$

The coordinates of the end points of the cable are:

$$\mathbf{A} = 0.8\hat{\mathbf{i}} + 0.6\hat{\mathbf{j}}$$

$$\mathbf{B} = 0.8\hat{\mathbf{j}} + 1.0\hat{\mathbf{k}}$$

A unit vector from A to B is:

$$\hat{\mathbf{t}} = \frac{\mathbf{B} - \mathbf{A}}{|\mathbf{B} - \mathbf{A}|}$$

$$\hat{\mathbf{t}} = -0.617\hat{\mathbf{i}} + 0.154\hat{\mathbf{j}} + 0.772\hat{\mathbf{k}}$$

A unit vector along the rod is $\hat{\mathbf{j}}$ and the vector equation of equilibrium is:

$$\mathbf{T} + \mathbf{N} + \mathbf{W} = 0$$

Taking dot product of this equation is taken with the unit vector along the rod, yields

$$\mathbf{T}\cdot\hat{\mathbf{j}} + \mathbf{N}\cdot\hat{\mathbf{j}} + \mathbf{W}\cdot\hat{\mathbf{j}} = 0$$

The normal force is perpendicular to the unit vector along the rod and the scalar equation yields:

$$T = \frac{-\mathbf{W}\cdot\hat{\mathbf{j}}}{\hat{\mathbf{t}}\cdot\hat{\mathbf{j}}} = 636\ \text{N}$$

The normal force can be obtained from the equilibrium equation:

$$\mathbf{N} = -(\mathbf{T} + \mathbf{W})$$

$$\mathbf{N} = 393\hat{\mathbf{i}} - 491\hat{\mathbf{k}}\ \text{N} \quad |\text{N}| = 629\ \text{N}$$

COMPUTATIONAL SOLUTION: SAMPLE PROBLEM 3.8

Enter the following in an m-file and safe it as SP38.m:

```
W=[0;-10*9.81;0];A=[0.8;0.6;0];B=[0;0.8;1.0];
t=(B-A)/norm(B-A); j=[0;1;0];
T=-W'*j/(t'*j),N=-(T'*t+W)
```

Then type SP38 after the prompt in the command window:

```
EDU>SM38
T =
   635.7607
N =
   392.4000
          0
  -490.5000
```

Sample Problem 3.9

The collars hang on the vertical frame composed of two smooth rods. If the mass of collar A is 8 kg and the mass of collar B is 4 kg, determine the equilibrium angle α and the tension in the cable between the collars.

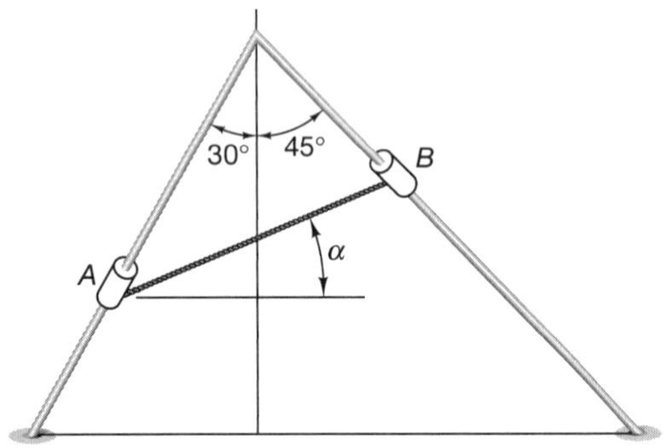

SOLUTION

Construct free-body diagrams for the two particles.

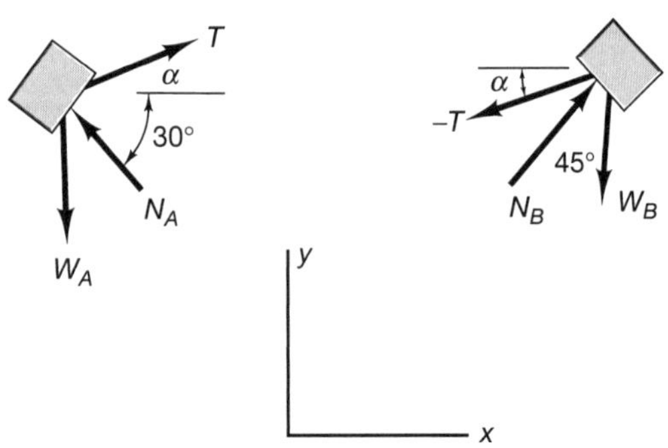

The equilibrium equation for collar A is

$$\mathbf{W_A} + \mathbf{N_A} + \mathbf{T} = 0$$

and for collar B is

$$\mathbf{W_B} + \mathbf{N_B} - \mathbf{T} = 0$$

Since all vector lie in a plane, the four scalar equations are:

$$T\cos\alpha - N_A\cos 30° = 0$$

$$T\sin\alpha + N_A\sin 30° - W_A = 0$$

$$-T\cos\alpha + N_B\sin 45° = 0$$

$$-T\sin\alpha + N_B\cos 45° - W_B = 0$$

Eliminating T and α yields

$$N_A \cos 30^\circ = N_B \sin 45^\circ \Rightarrow N_A = 0.816 N_B$$

$$N_A \sin 30^\circ + N_B \cos 45^\circ = 117.7$$

$$N_A = 86.1 \text{ N} \qquad N_B = 105.6 \text{ N}$$

From the first two equations:

$$T \cos\alpha = 74.6$$

$$T \sin\alpha = 35.6$$

$$\tan\alpha = 0.476$$

$$\alpha = 25^\circ$$

$$T = 82.3 \text{ N}$$

COMPUTATIONAL SOLUTION: SAMPLE PROBLEM 3.9

Type and save the following m-file as SamProb3_9.m

```
Function solution=SamProb3_9(x)
Na=x(1);  Nb=x(2);  T=x(3);  a=x(4);
% Given in the problem
Wa=200; Wb=300;
% Plan the equation
solution=[ T*cos(a)-Na*cos(30*pi/180);
         [ T*sin(a)+Na*sin(30*pi/180)-Wa;
          -T*cos(a)+Nb*sin(45*pi/180);
          -T*sin(a)+Nb*cos(45*pi/180)-Wb];

>>fsolve(@SamProb3_9,[100 100 75 30*pi/180])
ans =
    86.1770  105.5449  82.5979  0.4428
```

Sample Problem 3.10

A 10-kg collar at A and a 15-kg collar at B slide on smooth rods as shown. The collars are supported by a 2-m long cable. If the system is in equilibrium, determine the distance from the bottom of each rod to each collar and the tension in the cable.

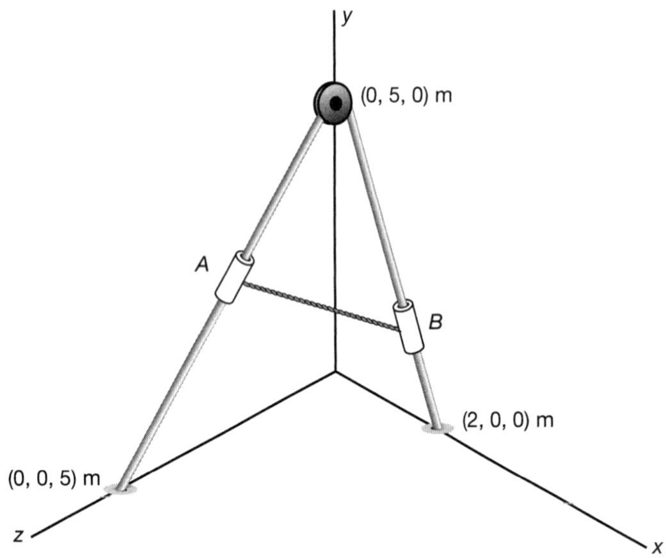

SOLUTION

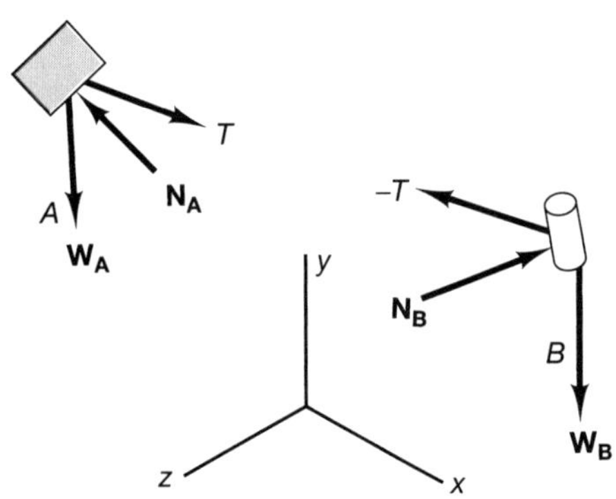

The equilibrium equation for collar *A* is

$$\mathbf{W_A} + \mathbf{N_A} + \mathbf{T} = 0$$

The equilibrium equation for collar *B* is

$$\mathbf{W_B} + \mathbf{N_B} - \mathbf{T} = 0$$

Now, construct unit vectors that run from the base of each rod directed toward the top of the rod.

$$\mathbf{L_A} = 5\hat{\mathbf{j}} - 5\hat{\mathbf{k}} \qquad \hat{\mathbf{l}}_A = 0.707\hat{\mathbf{j}} - 0.707\hat{\mathbf{k}}$$

$$\mathbf{L_B} = -2\hat{\mathbf{i}} + 5\hat{\mathbf{j}} \qquad \hat{\mathbf{l}}_B = -0.371\hat{\mathbf{i}} + 0.929\hat{\mathbf{j}}$$

The points A and B are located from the origin as

$$\mathbf{A} = 5\hat{\mathbf{k}} + a\hat{\mathbf{I}}_{\mathbf{A}}$$

$$\mathbf{B} = 2\hat{\mathbf{i}} + b\hat{\mathbf{I}}_{\mathbf{B}}$$

Where a and b are the distances from the bottom of the rod to the collar. A vector from collar A to collar B along the cable is

$$\mathbf{B} - \mathbf{A} = (2 - 0.371b)\hat{\mathbf{i}} + (0.929b - 0.707a)\hat{\mathbf{j}} - (5 - 0.707a)\hat{\mathbf{k}}$$

Since the cable has a length of 2 meters, a unit vector along $\mathbf{B} - \mathbf{A}$ is

$$\hat{\mathbf{t}} = \frac{\mathbf{B} - \mathbf{A}}{2}$$

The dot product may now be used to eliminate the normal forces from the equilibrium equations as these forces must be perpendicular to the rods.

$$T(\hat{\mathbf{I}}_{\mathbf{A}} \cdot \hat{\mathbf{t}}) - 10 \cdot 9.81(\hat{\mathbf{I}}_{\mathbf{A}} \cdot \hat{\mathbf{j}}) = 0$$

$$-T(\hat{\mathbf{I}}_{\mathbf{B}} \cdot \hat{\mathbf{t}}) - 15 \cdot 9.81(\hat{\mathbf{I}}_{\mathbf{B}} \cdot \hat{\mathbf{j}}) = 0$$

Expanding these two scalar equations yields:

$$T\,[0.707*0.5*(0.929b - 0.707a) + 0.707*0.5*(5 - 0.707a)] = 10*9.81*0.707$$

$$T\,[0.371*0.5*(2 - 0.371b) - 0.929*0.5*(0.929b - 0.707a)] = 15*9.81*0.929$$

These are two nonlinear equations for a, b and T. The third equation is obtained by observing that

$$(\mathbf{B} - \mathbf{A}) \cdot (\mathbf{B} - \mathbf{A}) = 4$$

$$(2 - 0.371b)^2 + (0.929b - 0.707a)^2 + (5 - 0.707a)^2 = 4$$

Computational software may be used to solve the three nonlinear equations, yielding

$$a = 4.85 \text{ m}$$

$$b = 2.84 \text{ m}$$

$$T = 252 \text{ N}$$

The normal forces may be found directly from the equilibrium equations.

$$\mathbf{N}_{\mathbf{A}} = -\mathbf{T} - \mathbf{W}_{\mathbf{A}}$$

$$\mathbf{N}_{\mathbf{B}} = \mathbf{T} - \mathbf{W}_{\mathbf{B}}$$

$$\mathbf{N}_{\mathbf{A}} = -119\hat{\mathbf{i}} + 198\hat{\mathbf{j}} + 198\hat{\mathbf{k}} \quad |\mathbf{N}_{\mathbf{A}}| = 304 \text{ N}$$

$$\mathbf{N}_{\mathbf{B}} = 119\hat{\mathbf{i}} + 48\hat{\mathbf{j}} - 198\hat{\mathbf{k}} \quad |\mathbf{N}_{\mathbf{B}}| = 236 \text{ N}$$

COMPUTATIONAL SOLUTION: SAMPLE PROBLEM 3.10

Save the following commands in an m-file named P3p10.m:

```
function solution=p3p10(x)

a = x(1);
b = x(2);
T = x(3);

% First equation
eq1 = T*(0.707*0.5*(0.929*b-0.707*a)+0.707*0.5*(5-0.707*a))-10*9.81*0.707;

% Second equation
eq2 = T*(0.371*0.5*(2-0.371*b)-0.929*0.5*(0.929*b-0.707*a))-15*9.81*0.929;

% Third equation
eq3 = (2-0.371*b)^2+(0.929*b-0.707*a)^2+(5-0.707*a)^2-4;

solution = [eq1;
            eq2;
            eq3];
```

Then type the following in the command window after the prompt to get the answer:

```
>> fsolve (@P3p10,[2 4 100],foptions)
ans =
   4.8474  2.8362  251.3612
```

Sample Problem 3.13

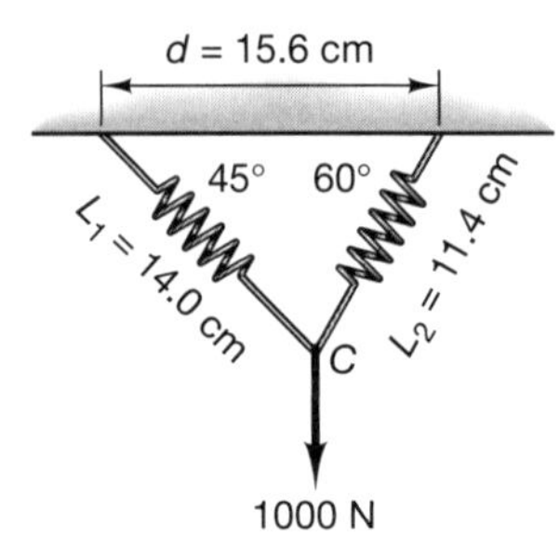

The spring system shown in the unstretched configuration at the right supports a vertical force **F** of 1000 N at point C. The lengths of the unstretched springs are $L_1 = 14$ cm and $L_2 = 11.4$ cm, and the spring constants are $k_1 = 2000$ N/cm and $k_2 = 1000$ N/cm; this is a very stiff system. Determine the tension in each spring. Solve the problem when the spring constants are reduced by a factor of 10—that is, when we have a "soft system."

After obtaining the equations to be solved, we next need to obtain an initial guess by solving the equilibrium equations using the undeformed configuration. The undeformed configuration results in linear equations that are solved by matrix methods. For this problem, the linear solution yields the following tensions and corresponding spring deflections:

$$T_1 = 518 \text{ N}, T_2 = 732 \text{ N}, d_1 = 0.259 \text{ cm, and } d_2 = 0.732 \text{ cm.}$$

These four values, along with two angles, will form the initial guess to be given the Newton–Raphson method of solving the system of nonlinear algebraic equations.

First create an m-file containing the equations to be solved. This file should be stored under the name 3pt13.m. Note that the file is a function file and calls for information outside of itself, so do not expect the file to execute until the appropriate input data are provided, namely the constants d, L_1, L_2, F, k_1, and k_2, as well as the vector p, which is the vector containing the initial guess. The file is as follows:

```
function q = 3pt13(p,d,L1,L2,F,k1,k2)
% names the function that contains the equations to be solved and indicates its
% dependency
alpha = p(1)*pi/180;    %this names the variables
beta = p(2)*pi/180;
T1 = p(3);
T2 = p(4);
del1 = p(5);
del2 = p(6);
q = zeros(6,1);    % this declares the size of q, initializes its values to zero,
                   %          and lists all the equations to be solved
q(1) = cos(alpha)-(d^2+(L1+del1)^2-(L2+del2)^2)/(2*d*(L1+del1));
q(2) = cos(beta)-(d^2+(L2+del2)^2-(L1+del1)^2)/(2*d*(L2+del2));
q(3) = T1 - k1*del1;
q(4) = T2 - k2*del2;
q(5) = T1*sin(alpha)+T2*sin(beta)-F;
q(6) = T1*cos(alpha)-T2*cos(beta);
```

Arranging the values to be provided from another file allows you to solve the same set of equations again for different parameter values, as is done later. Next, create an m-file to call the Newton–Raphson method (either fsolve or mnewt) and the m-file containing the equations to be solved:

```
x0 = [45; 60; 518;732; 0.259; 0.732]; % the vector of initial guesses to the
                                      % solution
d = 15.6;                             % set the values of the constant parameters
L1 = 14.0;
L2 = 11.4;
F = 1000;
k1 = 2000;
k2 = 1000;
x = fsolve('3pt13',x0,[],[],d,L1,L2,F,k1,k2); %if you have fsolve, use this
                                              %line if not, use
x = mnewt('3pt13', 10, x0, 0.001, 0.001, d,L1,L2,F,k1,k2);

fprintf('T1 = %8.3f\n',x(3));
fprintf('T2 = %8.3f\n',x(4));
fprintf('alpha = %8.3f degrees\n',x(1));
fprintf('beta = %8.3f degrees\n',x(2));
```

Now, executing this m-file reveals the solution of the equations, displaying those variables of interest:

```
EDU>3pt13
T1 =   519.411
T2 =   709.093
alpha =    47.538 degrees
beta =   60.393 degrees
```

The commands fsolve and mnewt require that several pieces of information be provided. The first piece of information is the name of the m-file containing the equations to be solved. In this case, the name is 3pt13, which is set in single quotes and followed by a comma. Next, mnewt requires the maximum number of trials allowed for the iteration (set at 10 here). This number is not required by the more sophisticated fsolve. The next entry is the vector containing the initial guess (x_0) set off by commas, and this is followed by tolerance for the x computation. (This is set to .001 in mnewt and to the default state in fsolve, which is denoted by [].) The y tolerance is then set to .001 in mnewt and to the default state in fsolve, again denoted by []. This is followed by a list of the constant parameters to be passed to the equations, each separated by commas.

The m-file used to call the command mnewt is followed by four fprintf commands. These are special output commands that allow the listing of name of the variable, followed by an equals sign and the computed value formatted to your specifications. In this case, the designation %8.3 tells MATLAB to print the value that follows in a field eight characters long and to a precision of three decimal places. The f that follows indicates fixed-point notation, and the \n instructs MATLAB to print on a new line. After the single quote and a comma, the name of the notation of the variable is given. So, for instance, if units such as degrees are required they can be included between the `f` notation and the `\n` notation, as indicated in the last two sentences. More information about this write command with formatted data command can be obtained by typing `help fprintf`.

The spring constants used in Sample Problem 3.13 were very stiff and the solution closely resembled that obtained by treating the cables as if they were rigid. It is useful to see the deformation and the tensions obtained if the stiffness of the system is reduced by a factor of 10. This solution is easily found by opening the m-file 3pt13.m and changing the values for k_1 and k_2 and to 200 N/cm and 100 N/cm, respectively. Then save and execute the file again. With these new values, the solution is as follows:

```
EDU>3pt13
T1 =  570.496
T2 =  556.555
alpha =   62.899 degrees
beta =   62.162 degrees
```

Sample Problem 3.14

Consider a scoreboard of mass 200 kg supported by three cables, as shown in the diagram at the right. Determine the tension in each of the cables if each cable has a spring constant equal to the product of its cross-sectional area and a material property (Young's modulus), divided by its length ($k = EA/L$). Assume that EA for each cable is equal to 1.2×10^5 N/m.

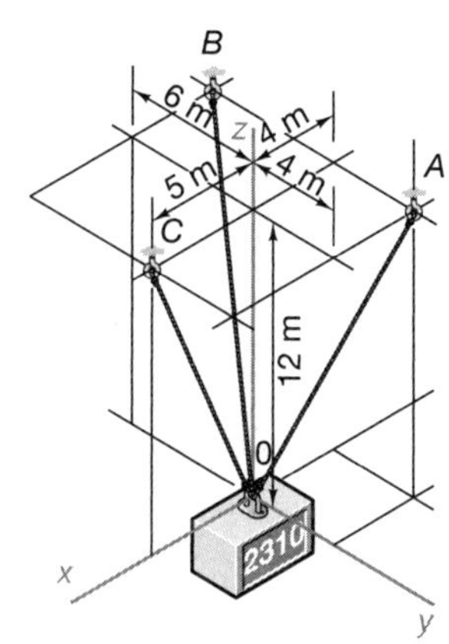

This sample problem, concerning a three-dimensional configuration of deformable cables, can also be solved using the commands `fsolve` and `mnewt`. However, we will solve it symbolically, because the equations have only quadratic nonlinearities. Solving the problem this way will give you some experience with using the symbolic commands in MATLAB as well. There are nine nonlinear equations to be solved in this case, and because we will use the symbolic commands, no initial guess is needed. Here the m-file that follows uses the `solve` command in MATLAB to solve the system of equations. First, the m-file is created by defining those variables that are to be treated as symbolic and for which we are solving. Next, all the constant values of the parameters are entered. These are the coordinates of the points of attachment of the cables. After that, the unstretched lengths of the cables are calculated in terms of these coordinates, followed by a calculation of the spring constants for each cable. Finally, the nine equations to be solved are entered. Note that they must be entered as an expression equal to zero, so that only one side of the equation is typed and named e.g., `S1=(xA-xo)^2+(yA-yo)^2+(zA-zo)^2-LA^2;`. The entire equation is followed by a semicolon, unless you want the equation to be recalculated in your command window. Once everything is entered, the solve command is used and returns the size of the solution for each variable when the m-file is run. After the file is run, the value of each variable is obtained by typing `S.` followed by the variable of interest. The code is as follows

```
syms TA TB TC xo yo zo LA LB LC            % establishes the list of symbolic
                                           % variables
xA=-4;yA=4;zA=12; EAA=1.2*10^5;            % enters the coordinates and cable
                                           % modulus
xB=-4;yB=-6;zB=12;EAB=1.2*10^5;
xC=5; yC=0; zC=12;EAC=1.2*10^5;
LAo=sqrt(xA^2 +yA^2 +zA^2);                % computes the unstretched lengths
LBo=sqrt(xB^2 +yB^2 +zB^2);
LCo=sqrt(xC^2 +yC^2 +zC^2);
kA=EAA/LAo;kB=EAB/LBo;kC=EAC/LCo;          % computes the stiffness of each cable
S1=(xA-xo)^2+(yA-yo)^2+(zA-zo)^2-LA^2;     % defines the equations to be solved
S2=(xB-xo)^2+(yB-yo)^2+(zB-zo)^2-LB^2;
S3=(xC-xo)^2+(yC-yo)^2+(zC-zo)^2-LC^2;
S4=TA-kA*(LA-LAo);
S5=TB-kB*(LB-LBo);
S6=TC-kC*(LC-LCo);
S7=(xA-xo)*TA/LA+(xB-xo)*TB/LB+(xC-xo)*TC/LC;
S8=(yA-yo)*TA/LA+(yB-yo)*TB/LB+(yC-yo)*TC/LC;
S9=(zA-zo)*TA/LA+(zB-zo)*TB/LB+(zC-zo)*TC/LC-1962;
S=solve(S1,S2,S3,S4,S5,S6,S7,S8,S9)       % calls for the solution stored in S
```

Next, save and execute this file (named `sp3pt14.m` here) to reveal the following

```
S =
     LA: [1x1 sym]
     LB: [1x1 sym]
     LC: [1x1 sym]
     TA: [1x1 sym]
     TB: [1x1 sym]
     TC: [1x1 sym]
     xo: [1x1 sym]
     yo: [1x1 sym]
     zo: [1x1 sym]
```

This states that only one solution was found. To obtain the symbolic values, type the following, which will return numerical answers if a symbolic solution is not found (note that the number of digits returned does not imply that the solution is accurate to the 24th decimal place):

```
EDU>Ta=S.TA,Tb=S.TB,Tc=S.TC,x0=S.xo,y0=S.yo,z0=S.zo
Ta =
722.57560128648612647111088693054
Tb =
515.79442266147544374676857773617
Tc =
935.54433771386047547392051344650
x0 =
-.38578153188817595185948732061855e-1
y0 =
-.21869345425439380825439705372046e-1
z0 =
-.93702511170957397125570551241694e-1
```

You should try this numerically using fsolve or mnewt to see whether, and if the solutions agree.

Sample Problem 3.15

A fourth cable D is added to the scoreboard in Sample Problem 3.14 for increased support. This cable is of the same material and cross-sectional area as the other three and is run 12 meters vertically to the ceiling in the undeformed condition. Determine the tension in the cables supporting the scoreboard.

The statically indeterminate Sample Problem 3.15 with 11 equations in 11 unknowns is solved using the fsolve command. If this command is not available, use the mnewt command from the appendix. First prepare an m-file containing the equations to be solved. Then save this file for use by calling fsolve from a second m-file. The code for the first m-file is as follows:

```
function q=sp3pt15(p,x,y,z,L,k,LOAD)
TA=p(1);TB=p(2);TC=p(3);TD=p(4);xo=p(5);yo=p(6);zo=p(7);   % name the variables
LA=p(8);LB=p(9);LC=p(10);LD=p(11);
xA  = x(1); xB  = x(2); xC  = x(3); xD  = x(4);             % name the constants
yA  = y(1); yB  = y(2); yC  = y(3); yD  = y(4);
zA  = z(1); zB  = z(2); zC  = z(3); zD  = z(4);
LAo = L(1); LBo = L(2); LCo = L(3); LDo = L(4);
kA  = k(1); kB  = k(2); kC  = k(3); kD  = k(4);
q=zeros(11,1);                                              % define the vector
                                                            % of equations
q(1)=(xA-xo)^2+(yA-yo)^2+(zA-zo)^2-LA^2;                    % list the equations
```

```
q(2)=(xB-xo)^2+(yB-yo)^2+(zB-zo)^2-LB^2;
q(3)=(xC-xo)^2+(yC-yo)^2+(zC-zo)^2-LC^2;
q(4)=(xD-xo)^2+(yD-yo)^2+(zD-zo)^2-LD^2;
q(5)=TA-kA*(LA-LAo);
q(6)=TB-kB*(LB-LBo);
q(7)=TC-kC*(LC-LCo);
q(8)=TD-kD*(LD-LDo);
q(9) =TA*(xA-xo)/LA+TB*(xB-xo)/LB+TC*(xC-xo)/LC+TD*(xD-xo)/LD;
q(10)=TA*(yA-yo)/LA+TB*(yB-yo)/LB+TC*(yC-yo)/LC+TD*(yD-yo)/LD;
q(11)=TA*(zA-zo)/LA+TB*(zB-zo)/LB+TC*(zC-zo)/LC+TD*(zD-zo)/LD-LOAD;
```

The following is the m-file that calls `fsolve` and the previous equations:

```
% Enter the coordinates of attachment points
%        A     B     C     D
x = [ -4    -4     5     4]';
y = [  4    -6     0    -4]';
z = [ 12    12    12    12]';
L = sqrt(x.^2 + y.^2 + z.^2);              % computes the unstretched cable lengths
% Enter the axial stiffness of cables
EA = [1.2e5   1.2e5   1.2e5   12e5]';
% calculate the spring constants
k = EA./L;
LOAD = 1962;                          % specify the load
% Place the initial guesses into a column vector in the following order:
%      TA    TB    TC     TD    xo    yo    zo    LA     LB    LC    LD
IC = [723;  509;  945;  100;   0;    0;    0;   L(1); L(2); L(3); L(4)];
soln = fsolve('sp3pt15',IC,OPTIONS,[],x,y,z,L,k,LOAD);             % call fsolve
fprintf("TA    = %8.3f\n',soln(1))
fprintf("TB    = %8.3f\n',soln(2))
fprintf("TC    = %8.3f\n',soln(3))
fprintf("TD    = %8.3f\n',soln(4))
fprintf("xo    = %8.3f\n',soln(5))
fprintf("yo    = %8.3f\n',soln(6))
fprintf("zo    = %8.3f\n',soln(7))
sp3pt15(soln,x,y,z,L,k,LOAD)
```

Saving and executing this file results in the following output:

```
TA    =  125.044
TB    =   89.374
TC    =  161.684
TD    = 1622.958
xo    =   -0.007
yo    =   -0.004
zo    =   -0.016
```

Notice that the addition of the fourth cable in the stated position causes that cable to carry almost all of the weight of the scoreboard. The remaining cables then serve mostly to stabilize the system. This is not a very reasonable design from the point of view of safety. A better design might be to move the fourth cable and attach it at $x_D = 4$ and $y_D = -4$, the rest of which is calculated next.

The foregoing file may be duplicated and renamed and the attachment coordinates changed to produce the following code:

```
% Enter the coordinates of the new attachment points
%       A    B    C    D
x = [ -4   -4    5    4]';
y = [  4   -6    0   -4]';
z = [ 12   12   12   12]';
L = sqrt(x.^2 + y.^2 + z.^2);               % computes the unstretched cable lengths
EA = [1.2e5   1.2e5   1.2e5   12e5]';       % Acomputes the axial stiffness of cables
k = EA./L;           % computes the spring constants
LOAD = 1962;         % enters the load
% Place the initial guesses into a column vector in the following order.
%      TA     TB     TC     TD     xo     yo     zo      LA    LB    LC    LD
IC = [723;  509;  945;  100;     0;     0;     0;   L(1); L(2); L(3); L(4)];
soln = fsolve("EA_3pt15_fcn',IC,OPTIONS,[],x,y,z,L,k,LOAD);
fprintf('TA    = %8.3f\n',soln(1))
fprintf('TB    = %8.3f\n',soln(2))
fprintf('TC    = %8.3f\n',soln(3))
fprintf('TD    = %8.3f\n',soln(4))
fprintf('xo    = %8.3f\n',soln(5))
fprintf('yo    = %8.3f\n',soln(6))
fprintf('zo    = %8.3f\n',soln(7))
sp3pt15(soln,x,y,z,L,k,LOAD)
```

Saving and executing this file yields the following solution:

```
TA    =  932.294
TB    =  192.870
TC    =  336.178
TD    =  708.868
xo    =    0.053
yo    =   -0.106
zo    =   -0.061
```

Note that the tensions for this configuration are more evently distributed and that all are less than 1000 N.

4

Rigid Bodies: Equivalent Force Systems

In this chapter, we consider rigid bodies for which the point of application of the force on the body is important to the equilibrium of the body. The turning effect of a force is defined as the moment of the force about a certain point. The moment is defined in terms of the vector or cross product of the position vector from that point to a point on the line of action of the force, i.e., $\mathbf{M} = \mathbf{r} \times \mathbf{F}$. MATLAB can be used to numerically and symbolically solve statics problems involving cross products as introduced in Computational Windows 2.4 and 2.6.

Vector or Cross Product Between Two Vectors

The MATLAB command for a cross product between two vectors `r` and `F` is `cross(r,F)`. The cross product calculation was used extensively at the end of Chapter 2. It is important to know how to compute cross products by hand when the vectors are fairly simple. But when the vectors are fully populated (i.e. no zero entries) using a computing device is likely to produce fewer errors, as well as giving quicker results. This was demonstrated in Computational Windows 2.5 and 2.6. The next example uses the vector product.

Sample Problem 4.4

Determine the moment of the force $P = 500$ N about each corner of the box shown in the diagram at the right.

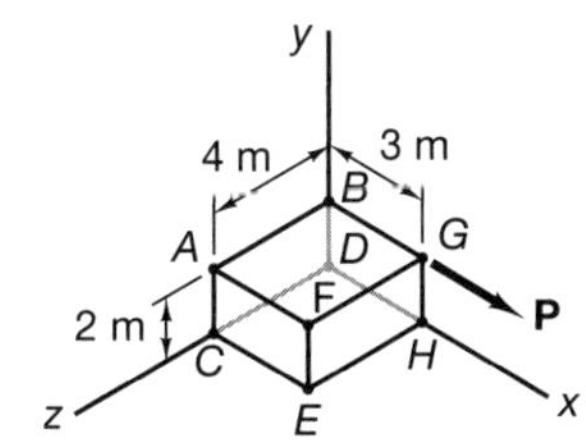

Sample Problem 4.4 will be solved using MATLAB. Note that MATLAB does not permit an operator (/) in the name of a quantity " so the vector **rB**/**A** is written simply as rBA. There is no convenient way to enter a subscript in MATLAB. One might be tempted to use r(BA) to refer to a subscript, but MATLAB will interpret this as the element of the vector r in the BAth position and will return an error if the value of BA is not assigned an integer between 1 and 3. We shall use the notation $A(i,j)$ and $A(i)$ to refer to the numerical values in the $(i\text{-}j)$ position of the matrix **A** and the ith position of the vector **A**, respectively. The solution to Sample Problem 4.4 is as follows:

COMPUTATIONAL SOLUTION: SAMPLE PROBLEM 4.4

```
EDU>P=[500;0;0];  rBA=[0;0;-4];  rBC=[0;2;-4];
EDU>Ma=cross(rBA,P), Mc=cross(rBC,P)
Ma =
          0
      -2000
          0
Mc =
          0
      -2000
      -1000
EDU>ma=norm(Ma), mb=morm(Mb)
ma =
       2000
mc =
   2.2361e+03
```

Sample Problem 4.5

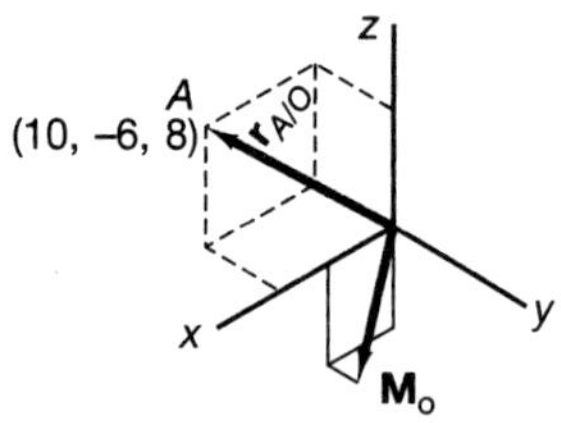

In the diagram at the right, determine the minimum force **F** that can be placed at point A, with coordinates $(10, -6, 8)$ m, to produce a moment at the origin given by

$$\mathbf{M}_o = 3000\hat{\mathbf{i}} + 1000\hat{\mathbf{j}} - 3000\hat{\mathbf{k}}\ \mathrm{N \cdot m}$$

The direct vector solution of Sample Problem 4.5 is easily obtained with any vector algebra program as follows:

COMPUTATIONAL SOLUTION: SAMPLE PROBLEM 4.5

% Define the moment $\mathbf{M}_0$ and the position vector $\mathbf{r}_{A/0}$ as column matrices:

```
EDU>MO=[3000;1000;-3000]; rAO=[10;-6;8];
EDU>Fp=-(cross(rAO,MO))/(dot(rAO,rAO))   % enter the direct vector formula
Fp =
   -50
  -270
  -140
```

Sample Problem 4.6

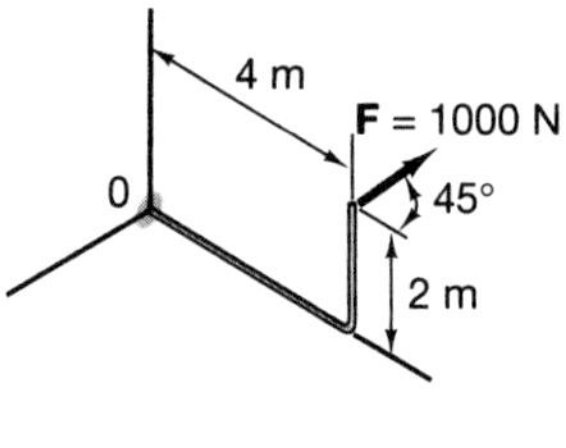

A crank is pinned at point O such that it is free to rotate about that point and is subjected to a 1000-N force, as shown in the figure at the right. Determine the moment at point O.

Although Varignon's theorem can be used to find the moment in many cases, especially when the dimensions are given in such a manner that perpendicular distances are obvious, vector algebra is usually the best way to write the moment equations:

COMPUTATIONAL SOLUTION: SAMPLE PROBLEM 4.6

```
EDU>Mo=cross([4;2;0],[70.7;70.7;0])
Mo =
         0
         0
  141.4000
```

The answer here is in Nm.

Sample Problem 4.7

A cable with a tension of 300 N is attached to the bar as shown. Calculate

i) the moment of the cable tension about point A using a position vector $\mathbf{r}_{B/A}$

ii) the moment of the cable tension about point A using a position vector $\mathbf{r}_{C/A}$

iii) the perpendicular position vector from A to the cable.

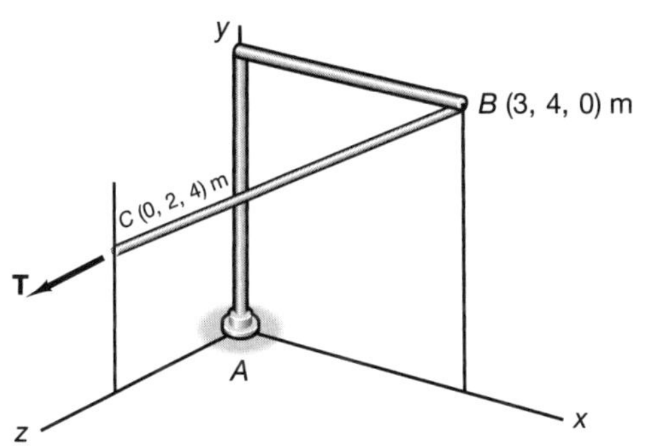

SOLUTION

The relative position vectors are:

$$\mathbf{r}_{B/A} = 3\hat{\mathbf{i}} + 4\hat{\mathbf{j}} \text{ m}$$

$$\mathbf{r}_{C/A} = 2\hat{\mathbf{j}} + 4\hat{\mathbf{k}} \text{ m}$$

$$\mathbf{r}_{C/B} = -3\hat{\mathbf{i}} - 2\hat{\mathbf{j}} + 4\hat{\mathbf{k}} \text{ m}$$

A unit vector along the cable is:

$$\hat{\mathbf{t}} = \frac{\mathbf{r}_{C/B}}{|\mathbf{r}_{C/B}|} = -0.557\hat{\mathbf{i}} - 0.371\hat{\mathbf{j}} + 0.743\hat{\mathbf{k}}$$

The tension vector in the cable is;

$$\mathbf{T} = T\hat{\mathbf{t}} = -167\hat{\mathbf{i}} - 111\hat{\mathbf{j}} + 223\hat{\mathbf{k}}\ \text{N}$$

The moment of the tension about point A is:

i) $\mathbf{M_A} = \mathbf{r_{B/A}} \times \mathbf{T} = 891\hat{\mathbf{i}} - 669\hat{\mathbf{j}} + 334\hat{\mathbf{k}}\ \text{Nm}$

ii) $\mathbf{M_B} = \mathbf{r_{C/A}} \times \mathbf{T} = 891\hat{\mathbf{i}} - 669\hat{\mathbf{j}} + 334\hat{\mathbf{k}}\ \text{Nm}$

The perpendicular position vector from A to the cable is defined by:

iii)
$$\mathbf{p} \times \mathbf{T} = \mathbf{M_A}$$
$$\mathbf{T} \times (\mathbf{p} \times \mathbf{T}) = \mathbf{T} \times \mathbf{M_A}$$
$$\mathbf{p}(\mathbf{T} * \mathbf{T}) - \mathbf{T}(\mathbf{T} * \mathbf{p}) = \mathbf{T} \times \mathbf{M_A}$$
$$\mathbf{p} = \frac{\mathbf{T} \times \mathbf{M_A}}{\mathbf{T} \cdot \mathbf{T}} = 1.24\hat{\mathbf{i}} + 2.83\hat{\mathbf{j}} + 2.35\hat{\mathbf{k}}\ \text{m}$$

$\mathbf{P} * \mathbf{T} = 0$ proves that $\mathbf{p}$ is perpendicular to $\mathbf{T}$.

COMPUTATIONAL SOLUTION: SAMPLE PROBLEM 4.7

Type the following in the an m-file named SP47.m

```
rBA=[3;4;0];rCA=[0;2;4];rCB=[-3;-2;4];%enter the relative position vectors
t=rCB/norm(rCB); %compute a unit vector along rCB
T=300*t;
MA=cross(rBA,t),MB=cross(rCA,T),p=cross(T,MA)./(T'*T)
```

Then save and execute to get:

```
MA =
   891.3376
  -668.5032
   334.2516
MB =
   891.3376
  -668.5032
   334.2516
P =
   1.2414
   2.8276
   2.3448
```

Sample Problem 4.8

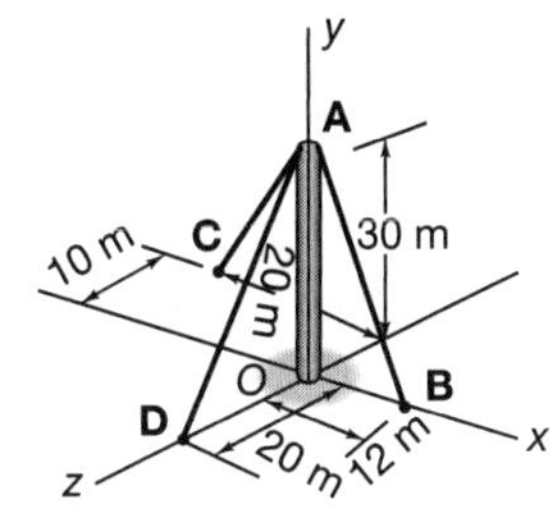

A pole is supported by three cables as shown in the diagram at the right. The tension in the cable AB is measured by a strain gage to be 3 kN. If the moment at the base of the pole O due to the tensions in the three cables is zero, determine the tension in the other two cables.

SOLUTION

Set up unit vectors along each of the three cables:

$$\mathbf{AB} = 12\hat{\mathbf{i}} - 30\hat{\mathbf{j}} \quad \hat{\mathbf{e}}_{\mathbf{AB}} = \frac{\mathbf{AB}}{|\mathbf{AB}|} \quad \hat{\mathbf{e}}_{\mathbf{AB}} = 0.371\hat{\mathbf{i}} - 0.928\hat{\mathbf{j}}$$

$$\mathbf{AC} = -20\hat{\mathbf{i}} - 30\hat{\mathbf{j}} - 10\hat{\mathbf{k}} \quad \hat{\mathbf{e}}_{\mathbf{AC}} = \frac{\mathbf{AC}}{|\mathbf{AC}|} \quad \hat{\mathbf{e}}_{\mathbf{AC}} = -0.535\hat{\mathbf{i}} - 0.802\hat{\mathbf{j}} - 0.267\hat{\mathbf{k}}$$

$$\mathbf{AD} = -30\hat{\mathbf{j}} + 20\hat{\mathbf{k}} \quad \hat{\mathbf{e}}_{\mathbf{AD}} = \frac{\mathbf{AD}}{|\mathbf{AD}|} \quad \hat{\mathbf{e}}_{\mathbf{AD}} = -0.832\hat{\mathbf{j}} + 0.555\hat{\mathbf{k}}$$

Establish a vector from the origin O to point A; $\mathbf{r}_{\mathbf{A/O}} = 30\hat{\mathbf{j}}$. The moment about the origin of the tension in each cable can now be formed as:

$$\mathbf{M}_{\mathbf{AB}} = \mathbf{r}_{\mathbf{A/O}} \times 3\hat{\mathbf{e}}_{\mathbf{AB}} = -33.42\hat{\mathbf{k}}$$

$$\mathbf{M}_{\mathbf{AC}} = \mathbf{r}_{\mathbf{A/O}} \times T_{AC}\hat{\mathbf{e}}_{\mathbf{AC}} = T_{AC}(-8.018\hat{\mathbf{i}} + 16.04\hat{\mathbf{k}})$$

$$\mathbf{M}_{\mathbf{AD}} = \mathbf{r}_{\mathbf{A/O}} \times T_{AD}\hat{\mathbf{e}}_{\mathbf{AD}} = T_{AD}16.64\hat{\mathbf{i}}$$

Setting the total moment at O to zero, yields:

$$T_{AC} = 2.08 \text{ kN}$$

$$T_{AD} = 1.00 \text{ kN}$$

This problem could also have been solved without writing the moment equations if you realize that if the sum of the tensions in the cable is in the y direction, the moment at the origin is zero.

$$3\hat{\mathbf{e}}_{\mathbf{AB}} + T_{AC}\hat{\mathbf{e}}_{\mathbf{AC}} + T_{AD}\hat{\mathbf{e}}_{\mathbf{AD}} = C\hat{\mathbf{j}}$$

COMPUTATIONAL SOLUTION: SAMPLE PROBLEM 4.8

```
EDU>C=[-8.018 16.04;16.14 0],MAB=[0;-33.42],
T=-inv(C)*MAB
C =
   -8.0180   16.0400
   16.1400         0
MAB =
         0
  -33.4200
T =
    2.0706
    1.0351
```

Sample Problem 4.9

It is known that the two cables produce a force perpendicular to the wall of 3.6 kN and a moment of magnitude of 10 kN m at point D. Determine the tension in the two cables.

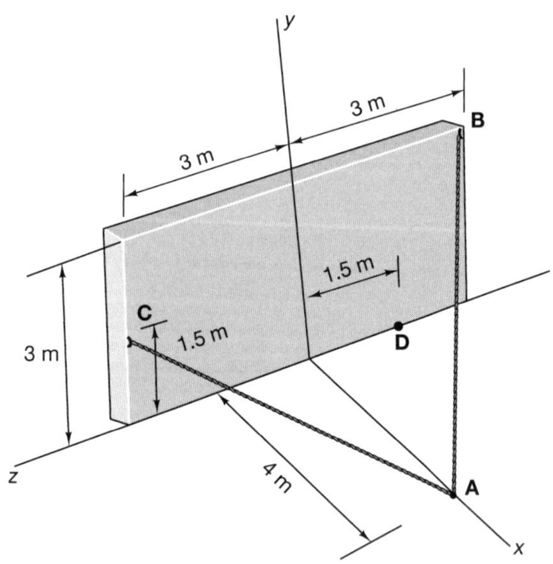

SOLUTION

Create relative position vectors from B to A and from C to A.

$$\mathbf{r}_{A/B} = 4\hat{\mathbf{i}} - 3\hat{\mathbf{j}} + 3\hat{\mathbf{k}}$$

$$\mathbf{r}_{A/C} = 4\hat{\mathbf{i}} - 1.5\hat{\mathbf{j}} - 3\hat{\mathbf{k}}$$

Form unit vectors along these two relative position vectors.

$$\hat{\mathbf{e}}_{\mathbf{A/B}} = 0.686\hat{\mathbf{i}} - 0.514\hat{\mathbf{j}} + 0.514\hat{\mathbf{k}}$$

$$\hat{\mathbf{e}}_{\mathbf{A/C}} = 0.766\hat{\mathbf{i}} - 0.287\hat{\mathbf{j}} - 0.575\hat{\mathbf{k}}$$

The tensions in the two cables can be written as:

$$\mathbf{T_B} = T_B\hat{\mathbf{e}}_{\mathbf{A/B}}$$

$$\mathbf{T_C} = T_C\hat{\mathbf{e}}_{\mathbf{A/C}}$$

The known fact that the cables produce a force perpendicular to the wall of 3.6 kN can be expressed mathematically as:

$$0.686T_B + 0.766T_C = 3.6$$

The second known fact that the magnitude of the moment at D is 15 kN m is more difficult to express mathematically. Form a position vector from D to A.

$$\mathbf{r_{A/D}} = 4\hat{\mathbf{i}} + 1.5\hat{\mathbf{k}}$$

We can now form the moment of each of the tensions about point D.

$$\mathbf{M_B} = \mathbf{r_{A/D}} \times T_B\hat{\mathbf{e}}_{\mathbf{A/B}}$$

$$\mathbf{M_C} = \mathbf{r_{A/D}} \times T_C\hat{\mathbf{e}}_{\mathbf{A/C}}$$

We know that the magnitude of the sum of these two vectors is 10 kN m, that is

$$|\mathbf{M_B} + \mathbf{M_C}| = 10$$

This is mathematically equivalent to writing

$$(\mathbf{M_B} \cdot \mathbf{M_B}) + 2(\mathbf{M_B} \cdot \mathbf{M_C}) + (\mathbf{M_C} \cdot \mathbf{M_C}) = 100$$

Expanding this equation yields

$$5.89T_B^2 - 1.7T_BT_C + 13.397T_C^2 - 100 = 0$$

This equation has the two unknown tensions as squares or a product and will have to be solved numerically with the first linear equation either by use of a calculator or computer. Write the two equations such that the tension T_C is a function of T_B

$$T_{C1}(T_B) = \frac{1}{0.766}(3.6 - 0.686T_B)$$

$$T_{C2}(T_B) = \frac{1}{2*13.397}\left[1.7T_B + \sqrt{1.7^2T_B^2 - 4*13.397*(5.89T_B^2 - 100)}\right]$$

We can obtain a solution graphically by plotting these two curves.

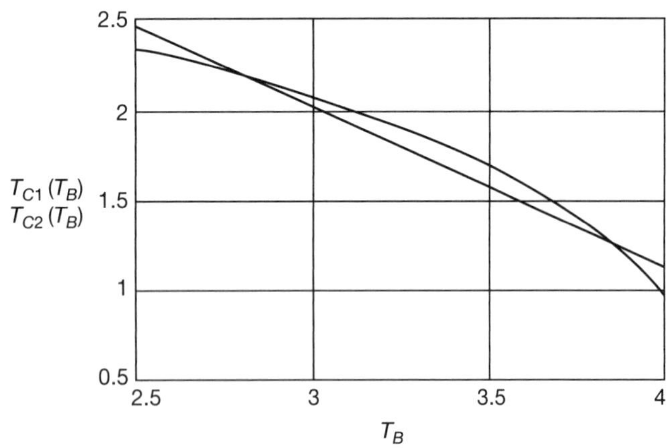

We can note that there are two solutions to the problem; one at each of the two intersections. The tensions in the cables are:

$$T_B = 2.8 \qquad 3.9 \text{ kN}$$

or

$$T_C = 2.2 \qquad 1.2 \text{ kN}$$

COMPUTATIONAL SOLUTION: SAMPLE PROBLEM 4.9

Save the following commands in an m-file named P4p9.m:

```
function solution=P4p9(x)

TB = x(1);
TC = x(2);

% First equation
eq1 = 0.686*TB+0.766*TC-3.6;

% Second equation
eq2 = 5.89*TB^2-1.77*TB*TC+13.397*TC^2-100;

solution = [eq1;
            eq2];
```

Then type the following the in the command window to obtain the solution:

```
>>fsolve (@P4p9,[4 1],foptions)
ans =
    3.8619    1.2411
```

Sample Problem 4.10

A tire lug wrench is used to remove the lug nuts on a tire. Compute the useful torque being exerted on the lug nut when a 200-N force is applied. (See figure below.)

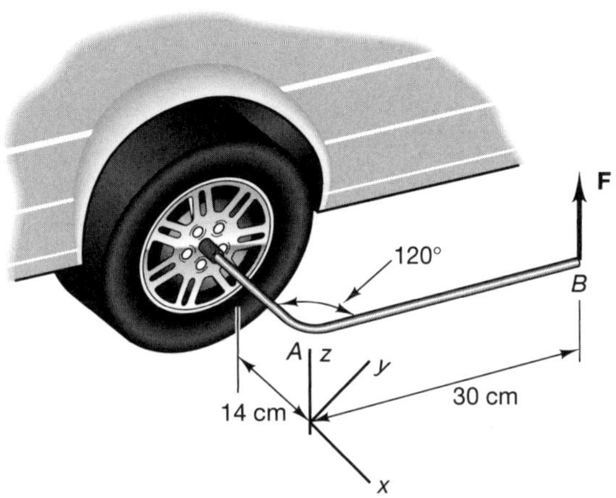

Calculating the component of a vector along an axis in space is an example of the use of the scalar triple product, as shown in Sample Problems 4.10 and 4.11. Note in these solutions that relative position vectors are written as **rBA** rather than $\mathbf{r}_{B/A}$. This is because MATLAB will not allow a mathematical operator in the name of a variable. The required vector algebra for Sample Problems 4.12 through 4.14 follows:

COMPUTATIONAL SOLUTION: SAMPLE PROBLEM 4.10

```
EDU> i=[1;0;0];   rBA=0.3*[sin(30*pi/180);cos(30*pi/180);0]; F=[0;0;200];
EDU>Mx=(dot(i,cross(rBA,F)))*i
Mx =
   51.9615
         0
         0
```

Sample Problem 4.11

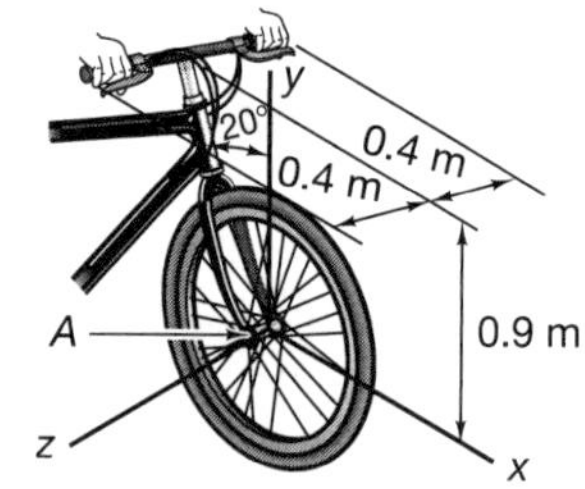

Suppose that $\mathbf{F}_r$ and $\mathbf{F}_l$ represent the forces applied to the right and left ends, respectively, of the handlebar of a bicycle, as shown in the diagram at the right. If the forces on the handlebar are $\mathbf{F}_r = (-80\hat{\mathbf{i}} - 80\hat{\mathbf{j}})$ N and $\mathbf{F}_l = (-80\hat{\mathbf{i}} - 80\hat{\mathbf{j}} + 40\hat{\mathbf{k}})$ N, determine the moment exerted by the rider at the center of the wheel axis. Then determine the component of the moment about the y-axis.

COMPUTATIONAL SOLUTION: SAMPLE PROBLEM 4.11

```
EDU>rrA=[-0.9*sin(20*pi/180); 0.9*cos(20*pi/180); 0.4]; Fr=[-80;-80;0];
EDU>rlA=[-0.9*sin(20*pi/180); 0.9*cos(20*pi/180); -0.4]; Fl=[-80; -80; 40];
EDU>j=[0;1;0];   % defines the unit vector along the y-axis
EDU>MA=cross(rrA,Fr)+cross(rlA,Fl)
MA =
   33.8289
   12.3127
  184.5666
EDU>My=dot(MA,j)
My =
   12.3127
```

Sample Problem 4.12

The foundation of a building constructed to sit in water is extended to spread the weight of the building over a large surface area. This configuration is called a “raft” or “mat” foundation. If the loading on the foundation is as shown in the following figure, calculate the moment of the couples acting on it.

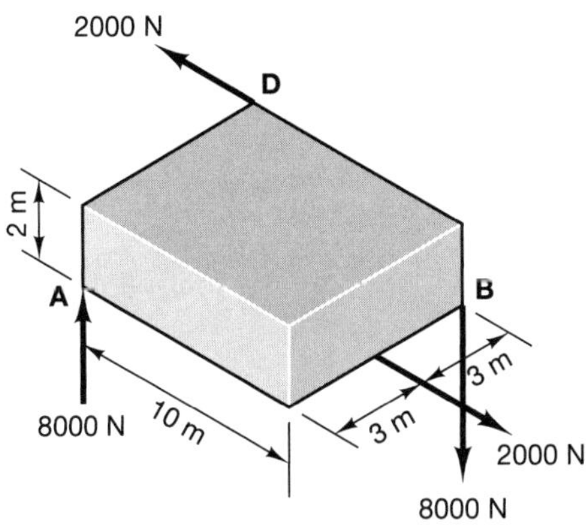

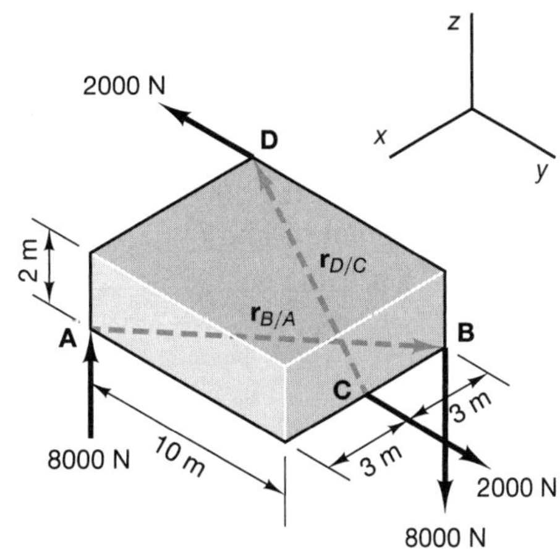

COMPUTATIONAL SOLUTION: SAMPLE PROBLEM 4.12

First enter the required vectors, noting that the forces are in newtons and the positions in meters.

```
EDU>rBA= [-6;10; 0]; rDC= [-3;-10; 2]; FB= [0; 0;-8000]; FD= [0;-2000; 0];
EDU>CAB=cross(rBA,FB), CCD=cross(rDC,FD), M=CAB+CCD    % compute the needed
                                                       % products

CAB =
      -80000
      -48000
           0
CCD =
        4000
           0
        6000
M =
      -76000
      -48000
        6000
```

Sample Problem 4.13

In the diagram at the right, determine the equivalent force systems consisting of a force and the moment of a couple if the 1000-N load applied at point A is moved to point B and then to points C and D.

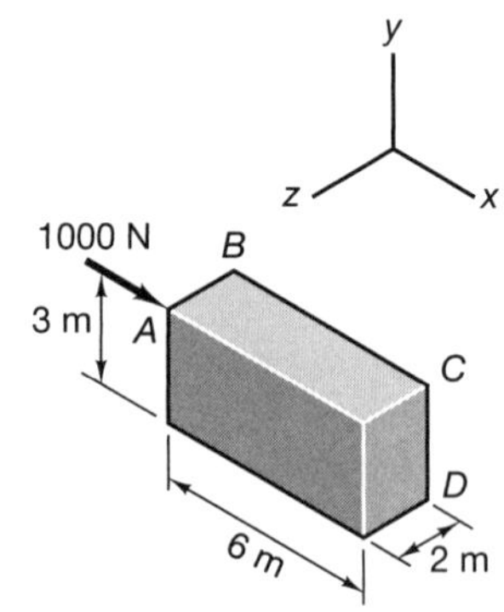


```
EDU>rAB=[0;0;2];FA=[1000;0;0];CB=cross(rAB,FA)
CB =
         0
      2000
         0
EDU>rAC=[-6;0;2];CC=cross(rAC,FA)
CC =
         0
      2000
         0
EDU>rAD=[-6;3;2];CD=cross(rAD,FA)
CD =
         0
      2000
     -3000
```

Sample Problem 4.14

A coplanar force system consists of the following forces with position vectors from the origin, as given:

$$\mathbf{F}_1 = 100\hat{\mathbf{i}} + 100\hat{\mathbf{j}}\ (\text{N}) \qquad \mathbf{r}_{1/O} = 2\hat{\mathbf{i}} + 3\hat{\mathbf{j}}\ (\text{m})$$

$$\mathbf{F}_2 = 50\hat{\mathbf{j}} \qquad \mathbf{r}_{2/O} = -10\hat{\mathbf{i}} + 2\hat{\mathbf{j}}$$

$$\mathbf{F}_3 = 300\hat{\mathbf{i}} - 450\hat{\mathbf{j}} \qquad \mathbf{r}_{3/O} = 4\hat{\mathbf{i}} - 4\hat{\mathbf{j}}$$

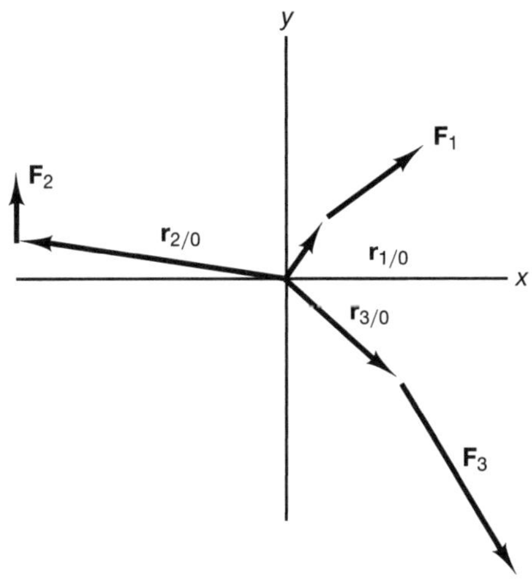

Determine the equivalent force system consisting of a single resultant, and determine the line of action of the resultant in space.

COMPUTATIONAL SOLUTION: SAMPLE PROBLEM 4.14

Define the force and position vectors of the coplanar force system, and enter them.

```
EDU>F1= [100;100; 0]; F2= [0; 50; 0]; F3= [300;-450; 0];     % enters the force
                                                              % vectors
EDU>r1= [2; 3; 0]; r2= [-10; 2; 0]; r3= [4;-4; 0];           % enters the position
                                                              % vectors
EDU>C= cross(r1,F1)+cross(r2,F2)+cross(r3,F3), R= F1+ F2+ F3
C =
          0
          0
      -1200
R =
    400
   -300
      0
EDU>r=(cross(R,C))/(dot(R,R))
r =
     1.4400
     1.9200
          0
```

Sample Problem 4.16

In biomechanical studies of human motion, a dynamometer in the form of a force plate is placed in the floor to measure the forces between the foot and the ground when an individual walks. The force plate uses strain gages to measure the three components of force and the three components of the moment about the center of the instrument, which is 40 mm below the surface of the force plate, as shown in the following diagram: Suppose the force plate measures a resultant force and a moment of

$$\mathbf{R} = 50\hat{\mathbf{i}} + 150\hat{\mathbf{j}} + 800\hat{\mathbf{k}} \text{ (N)}$$

$$\mathbf{M}_O = 80\hat{\mathbf{i}} + 10\hat{\mathbf{j}} + 10\hat{\mathbf{k}} \text{ (N}\cdot\text{m)}$$

First, determine the wrench force and moment (called the ground reaction torque in this application). Then determine the intercept of the wrench axis with the surface of the force plate. (This intercept point, called the "center of pressure" or "center of force," is the point at which the force is concentrated between the foot and the plate.)

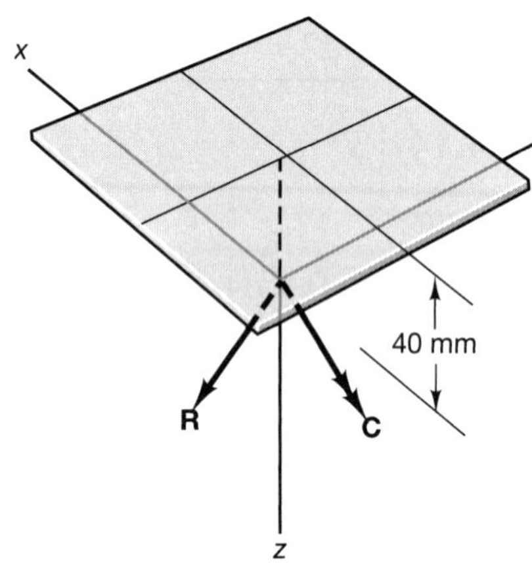

The calculation of the equivalent wrench for a given force system involves many vector operations that are most easily done using MATLAB, as shown in this sample problem and in Sample Problem 4.17.

COMPUTATIONAL SOLUTION: SAMPLE PROBLEM 4.16

```
EDU>R= [50;150; 800]; Mo= [80; 10; 10];   % enter the force, and moment at the
                                          % instruments center
EDU>eR=R/norm(R)                          % compute the unit vector along R
eR =
    0.0613
    0.1839
    0.9810
EDU>T= dot(Mo,eR)*eR   % enter the formula for the torque parallel to the vector R
T =
    1.0150
    3.0451
   16.2406
EDU>Mp =Mo-T   % enter the formula for the perpendicular component of the couple
Mp =
   78.9850
    6.9549
   -6.2406
EDU>po= (cross(R,Mp))/(dot(R,R))  % Compute the perpendicular vector from the
                                  % orgin to the wrench axis

po =
   -0.0098
    0.0955
   -0.0173
```

Note that P_0 is not the vector to the surface of the plate, but a vector perpendicular to the wrench axis.

The intercept with the surface of the force plate **r** can now be found by moving from the perpendicular intercept point along the wrench axis to the surface of the plate a distance d.

$$\mathbf{r} = \mathbf{p} + d\hat{\mathbf{e}}_{\mathbf{R}}$$

$$x\hat{\mathbf{i}} + y\hat{\mathbf{j}} - 0.04\hat{\mathbf{k}} = (-0.01\hat{\mathbf{i}} + 0.095\hat{\mathbf{j}} - 0.017\hat{\mathbf{k}}) + d(0.061\hat{\mathbf{i}} + 0.184\hat{\mathbf{j}} + 0.981\hat{\mathbf{k}})$$

$$d = \frac{-0.04 + 0.017}{0.981} \quad d = -0.023$$

$$x = -0.01 + d{*}0.061$$

$$y = 0.095 + d\cdot 0.184$$

$$x = -0.011$$

$$y = 0.091$$

This agrees with the solution presented in the text.

Sample Problem 4.17

It is known that only a moment in the z-direction can be generated on the surface of a force plate in a gait analysis laboratory. The force and moment at the instrument's center are, respectively,

$$\mathbf{R} = 50\hat{\mathbf{i}} - 150\hat{\mathbf{j}} + 450\hat{\mathbf{k}} \text{ N}$$

$$\mathbf{M}_O = 13.5\hat{\mathbf{i}} + 9\hat{\mathbf{j}} - 6\hat{\mathbf{k}} \text{ N} \cdot \text{m}$$

Determine an equivalent force system composed of the resultant force and a moment in the z-direction. Also, determine a perpendicular vector from the center of the instrument to the line of action of the resultant in the equivalent system.

COMPUTATIONAL SOLUTION: SAMPLE PROBLEM 4.17

```
EDU>R= [50; -150; 450]; Mo= [13.5; 9; -6]; n= [0; 0; 1]; % enter vectors, units
                                                          % are Nm and N.
EDU>T=dot(R,Mo)/dot(R,n)
T =
   -7.5000
EDU>po= cross(R,Mo-T*n)/dot(R,R)
po =
   -0.0188
    0.0264
    0.0109
```

5

Distributed Forces: Centroids and Center of Gravity

Numerical and Symbolic Integration

To determine the centroid of a line, an area, or a volume, integrals must be evaluated. Integration is a standard topic in introductory calculus courses. In this chapter, we show how integrals may be evaluated either numerically or symbolically using MATLAB. The *Student Edition of MATLAB* presents numerical integration, in Section 16.4 (page 173) and symbolic integration in Section 20.15 (page 274). First consider using symbolic integration, as illustrated in the Computational Window 5.1. Recall from Computational Window 2.4 that symbolic manipulations require that you first define all the symbols to be used as symbolic quantities.

COMPUTATIONAL WINDOW 5.1

% The centroid for the triangle shown in the figure is evaluated
% symbolically using single integration as follows:

```
EDU>syms h b y              % declare that h, b, and y are symbolic values
EDU>f=2*(h-y)*y/h^2         % define the symbolic function to be integrated
f =
(2*h-2*y)*y/h^2
EDU>int(f,y,0,h)            % integrate the symbolic function f with respect to y,
                              % from 0 to h
ans =
                                      1/3*h
```

Note that MATLAB reads the answer as 1/3 times h, not $1/(3h)$, as it may appear. MATLAB performs operations from left to right. Next, the integral is evaluated symbolically using double integration:

```
EDU>syms a b h x y    % declare that h, b, x, and y are symbolic values
EDU>f1=y;             % define the fuction to be integrated over x
EDU>f2=int(f1,x,a*y/h,b-((b-a)/h)*y);     % integrate f1 over x from ay/h to
                                          % b-((b-a)/h)y
EDU>yc=(2/(b*h))*int(f2,y,0,h)       % now integrate f2 over y from 0 to h and
                                     % multiply by 2/bh
yc =
1/3*h
EDU>pretty(yc)        % this command makes the answer look pretty!
  1/3 h
```

Observe that, although the computational software replaces tables of integrals, the integrals must be correctly formulated and the limits of integration chosen correctly before symbolic integration can be used successfully. This capability, of course, does not replace the need to understand calculus: In fact, a clear understanding of double integration is required to perform centroid integrations in MATLAB and to choose the correct limits of integration and integrands. Recall that `pi` is the symbol used in MATLAB to denote π, and it is automatically treated as a symbolic quantity.

Sample Problem 5.1

Find the centroid of the semicircular area shown at the right.

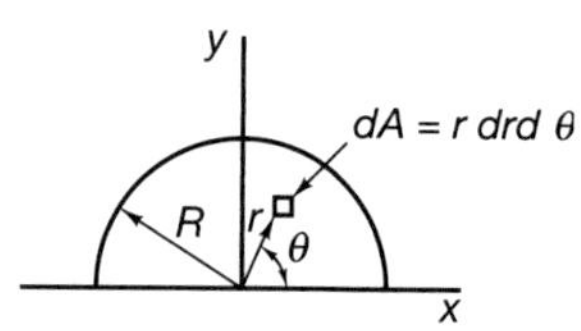

The solution to this problem via symbolic manipulation is given in the next window. The centroid of a semicircular area is computed as follows:

COMPUTATIONAL SOLUTION: SAMPLE PROBLEM 5.1

```
EDU>syms R r t           % define R, r and t as symbolic quantities (using t for
                         % theta)
EDU>f1=r;                % define the first integrand
EDU>f2=int(f1,r,0,R)     % integrate f1 over t from 0 to R and name it f2
f2 =
1/2*R^2
EDU>A=int(f2,t,0,pi)     % calculate the area by integrating f2 over t from 0 to pi
A =
1/2*pi*R^2
EDU>pretty(A)            % make the result more legible
                                        2
                                 1/2 pi R
EDU>f3=r^2;              % define the first integrand
EDU>f4=int(f3,r,0,R)     % integrate f3 over r from 0 to R
f4 =
1/3*R^3
EDU>y=(2/(pi*R^2))*int(f4*sin(t),t,0,pi)    % define y in terms of the integral
                                            % of f4 over t from 0 to pi.
y =
4/3/pi*R
EDU>pretty(y)            % make the result more legible
                                        R
                                 4/3 ----
                                       pi
```

Next, we introduce numerical integration. The previous sample problems have all been integrated symbolically, and as you may know from using a table of integrals, symbolic, or closed-form, solutions do not always exist. One can, however, always integrate numerically by various methods. MATLAB provides three different commands for numerically integration (`trapz`, `quad`, and `quad8`), of varying degrees of accuracy and sophistication. The theory of the numerical evaluation of an integral can be found in your calculus textbook and is discussed in detail in other texts on numerical methods. Basically, the methods of numerical integration estimate the area under the curve of a plot of the function to be integrated. The command `trapz` can be used to integrate a function defined in the command window, while the more acurate `quad` and `quad8` require you to first set up an m-file containing the function that you want to integrate (called the integrand). After setting up this file, you simply call it from one of the two integration commands. The fastest, but least accurate, method is trapz, which uses a step size that you choose. The quad command uses an automated step size, it takes longer, but gives more accurate results. The quad8 command is the most accurate method, but takes the longest to compute. For most integrals encountered in statics, quad8 is appropriate, as computational time is not usually an issue.

Sample Problem 5.3

Determine the centroid of the family of curves $y = x^n$.

Several integrals are required to compute centroids. We first compute the integral L for the length of the curve for the value $n = 1$. The first step is to create an m-file containing the integrand by choosing **File** and **New M-file** from the tool bar and typing the following:

```
function F=cent1(x)
n=1;
F=sqrt(n^2*x.^(2*(n-1))+1);  % the integrand or function to be integrated
```

This m-file must be saved and named cent1.m, of course. Next, type the following in the command workspace to perform the desire integration.

```
EDU>quad('cent1',0,1)     % integrate the function stored in cent1.m from 0 to 1
ans =
  1.4142
```

This procedure can be repeated to generate all of the integrals needed to compute the centroids. Naturally, the m-files must be created for each function to be integrated. We shall consider the case where $n = 5$, the other cases can be performed yourself following the same steps. First the m-file cent1, previously entered, is copied, the value of n is changed to 5, and the new file is named cent5. Next, create two more m-files to compute the integral for the x value of the centroid and the y value of the centriod by entering the following code:

```
function  f=xcent5(x)
n=5;
f=sqrt((n^2)*x.^(2*(n-1))+1).*x;
```

Note that the dot (*) multiplication and dot exponential are used because the functional form indicated that the function is to be evaluated element by element. This is also true in defining the integrand for the y centroid as follows:

```
function  f=ycent5(x)
n=5;
f=sqrt((n^2)*x.^(2*(n-1))+1).*x.^n;
```

With these three m-files (cent5, xcent5, and ycent5) defined, the integrals can be evaluated and the centroid computed, as indicated in the following window:

COMPUTATIONAL SOLUTION : SAMPLE PROBLEM 5.3

```
EDU>xc5=quad8('xcent5',0,1)/quad8('cent5',0,1);
EDU>yc5=quad8('ycent5',0,1)/quad8('cent5',0,1);
EDU>xc5,yc5
xc5 =
    0.6460
yc5 =
    0.3261
```

MATLAB as a Programming Language

Next, we show how to calculate and plot the value of a function involving an integral, as illustrated in the *Statics* text. This will require a little programming in MATLAB, a skill that will be valuable in classes you take later. One of MATLAB's strengths is that it can be used both as a calculator and as a serious programming language. Here we need to introduce the concept of *control flow,* or *looping,* in MATLAB. Looping allows you to repeat steps, to let past computations influence future ones, and to make a decision based on a past computation. As an example of using a for loop, consider generating a conversion table from degrees to radians, as illustrated in the following window:

```
EDU>for deg=1:6          % call the for command and let deg range over the integers
  1,2,3,4,5 &6
rad(deg)=deg*pi/180;     % define a row vector rad to have elements equal to the
                         % radian version of 1, 2,...
end                      % this denotes the end of the loop that will go 6 times.
EDU>rad                  % display the result
rad =
    0.0175    0.0349    0.0524    0.0698    0.0873    0.1047
```

Now we use quad8 integration and a for loop to integrate and plot the position of the centriod for various values of s. This requires the quad8 integration command, which in turn requires that m-files be created and saved which contain the functions to be integrated. Since the integration of the function z is trivial, the analytical value of its integral is used. The following two m-files, named f1.m and f2.m, respectively, set up the functions to be integrated:

```
function y=f1(zeta,R,beta)  % defines the integrand for the x value in terms of
                             % the values zeta, R, and beta which must be supplied
                             % by the command or m-file that calls this m-file
r = R*cos(beta);            % define r in terms of the values R and beta
theta = zeta/r;             % define theta in terms of r and zeta
y = cos(theta)*cos(beta);   % define the integrand
```

```
function y=f2(zeta,R,beta)  % defines the integrand for the y value in terms of
                             % the values zeta, R, and beta which must be supplied
                             % by the command or m-file that calls this m-file
r = R*cos(beta);            % define r in terms of the values R and beta
theta = zeta/r;             % define theta in terms of r and zeta
y = sin(theta)*cos(beta);   % define the integrand
```

With the two integrands defined and saved in m-files, the following m-file program will compute and plot $x(s)$, $y(s)$ and $z(s)$:

COMPUTATIONAL WINDOW 5.2

```
% A helical spiral of radius r and pitch p is shown using equation (5.24)
R = 2;                      % define the value of R
beta = pi/6;                % define the value of beta
x0 = 0;                     % set the initial value of x
y0 = -1.5;                  % set the initial value of y
z0 = 0;                     % set the initial value of z
s=0:.1:12;                  % define range of the variable upper limit of
                            % integration
n=length(s);                % number of integrations = n
% numerical integration; repeated for each individual value of the upper
% limit s(i) using a for loop
for i = 1:n;                % we will do this n times, for each integer from 1
                            % to n
 xs(i) = x0 + quad8('f1',0,s(i),[],[],R,beta);    % integrate f1 from 0 to s(i)
                                                  % and store it in xs(i)
 ys(i) = y0 + quad8('f2',0,s(i),[],[],R,beta);    % integrate f2 from 0 to s(i)
                                                  % and store it in ys(i)
```

```
end;                            % denotes end of loop
zs = z0 + sin(beta).*s;         % evaluate the integrated form of z(s)
plot(ys,xs,'y-',ys,zs,'y+')     % plot the functions
ylabel('x(s)- & z(s)+')         % label the y axis
xlabel('y(s)')                  % label the x axis
```

This m-file is saved under the name sp5pt3. Thus, typing sp5pt3 after the EDU> prompt will result in the following plot.

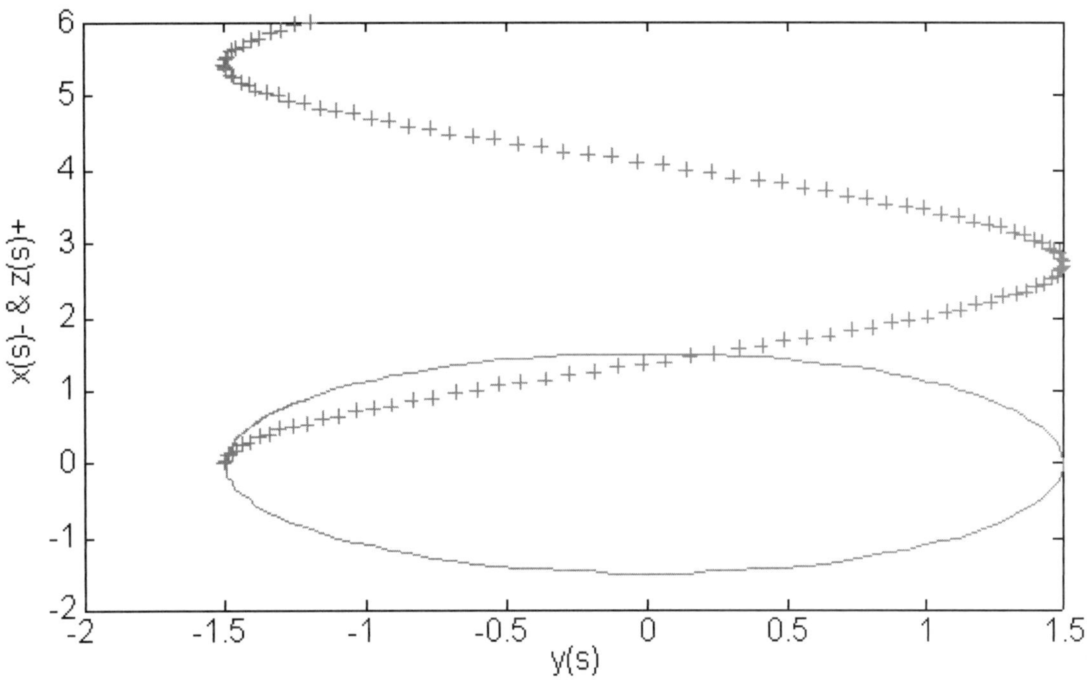

Next, we would like to calculate the length and centroid for the interger values of n from 1 to 5. First we create three m-files, containing the integrand for the length, the x centroid and the y centroid, respectively. Then we write an m-file that calls these functions and integrates them for each value of n. The m-file for the length is as follows:

```
function y=nIL(x,n)     % defines the length integrand
y=sqrt((n^2)*(x.^(2*n-2))+1);
```

These steps are saved in a file named `nIL.m`. The m-file for the integrand for the *x* centroid is illustrated in the following window:

```
function y=nIXC(x,n)     % define the integrand for the x centroid as a function
                         % of x and n;
y=sqrt((n^2)*(x.^(2*n-2))+1).*x;
```

Then these steps are saved in a file named nIXC. The m-file for the integrand for the *y* centroid is defined as follows:

```
function y=nIYC(x,n)     % defines the integrand for the y centriod as a function
                         % of x and n;
y=sqrt((n^2)*(x.^(2*n-2))+1).*x.^n;
```

Now these steps are saved in a file named `nIYC.m`. The m-file for computing the integrals of the functions is shown in the following window:

```
% calculate the various intergrals needed for the centroids for n = 1 through 5
for n=1:5
L=quad8('nIL',0,1,[],[],n);         % integrate nIL from 0 to 1 over x for the
                                    % given value of n
xc=quad8('nIXC',0,1,[],[],n)./L;    % integrate nIXC from 0 to 1 over x for the
                                    % given values of n and divide by the number L
yc=quad8('nIYC',0,1,[],[],n)./L;    % integrate nIYC from 0 to 1 over x for the
                                    % given values of n and divide by the number L
A(:,n)=[n L xc yc]';                % write the answers into the matrix A
end                                 % end the loop
A                                   % display the matrix A containing the
                                    % centroids for each n
```

This file is saved under the name `sp5pt3.m`. Thus, typing `sp5pt3` after the EDU> prompt displays the following data, which can be read as values of *n* across the top, followed by a row of values of length *L*, a row of values of the *x* centroid, and a row of values of the *y* centroid:

```
A =
      1.0000     2.0000     3.0000     4.0000     5.0000
      1.4142     1.4789     1.5479     1.6002     1.6406
      0.5000     0.5736     0.6086     0.6306     0.6460
      0.5000     0.4100     0.3664     0.3418     0.3261
```

It is instructive to note that a centroid does not necessarily lie on the object. To see this, the curve $y = x^5$ is plotted, and the centroid for this curve is labeled on the plot as directed in the following m-file:

```
x=0:.1:1;   % set the range of values for x
y=x.^5;     % define the function y = x^5 using the element-by-element exponent
plot(x,y)   % plot y versus x
xlabel('x')% the following provide the appropriate labels
ylabel('y = x^5')
text(0.646,0.326,'+')  % using the location of the centroid from the preceding
                       % solution, mark it text(0.6,.42,'centroid at')
text(0.6,.38,'0.646, 0.326')
```

Executing this m-file yields the following plot, which illustrates that the centroid does not lie on the line:

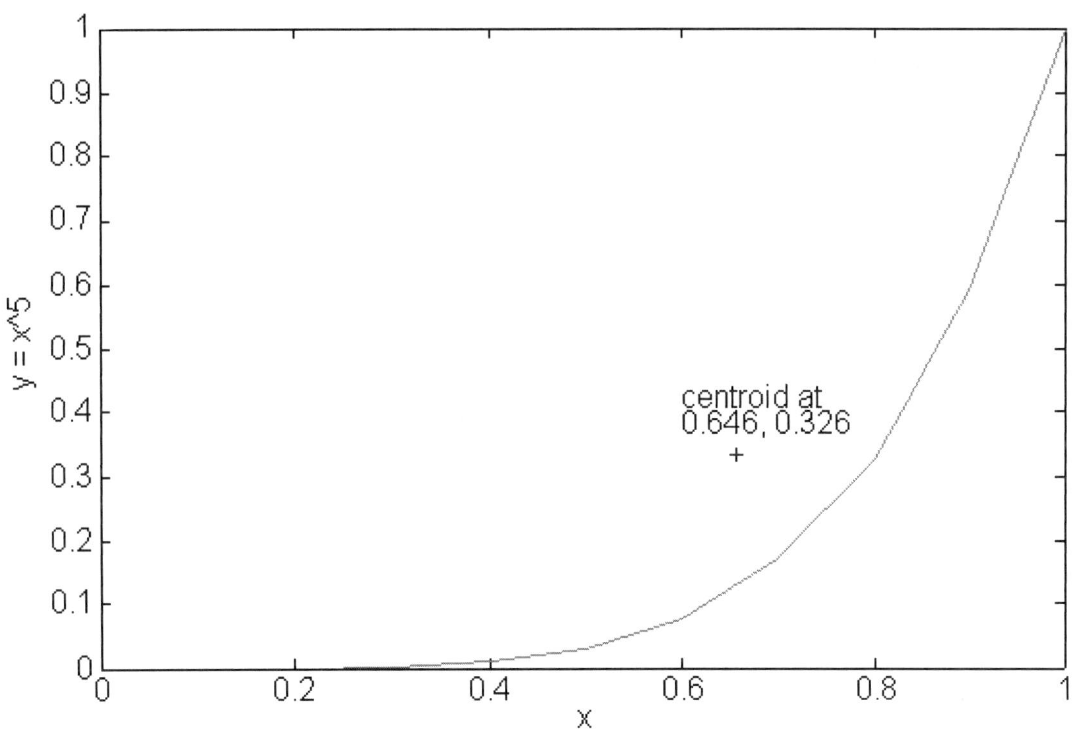

6

Equilibrium of Rigid Bodies

Several sample problems from Chapter 6 of *Statics* are solved by MATLAB as examples of how this software may be used to reduce the computational burden. For example, Sample Problem 6.4 requires the solution of a transcendental equation. This problem can be solved using MATLAB by first graphing Eq. (SP6.4.6) and then finding initial estimates of the roots. The `fsolve` command is then used to determine the exact roots. This procedure is identical to that described in Chapter 1 under the section headed "Solving an Algebraic Equation." In this regard, although we usually think of roots of quadratic or cubic equations, fsolve can be used to solve for the roots of other transcendental equations. Since this command is of an iterative nature, that is, starting with an initial guess and using a Newtonian method to seek the roots, an initial starting point is required. The initial guess is obtained by plotting the function and noting where it crosses zero.

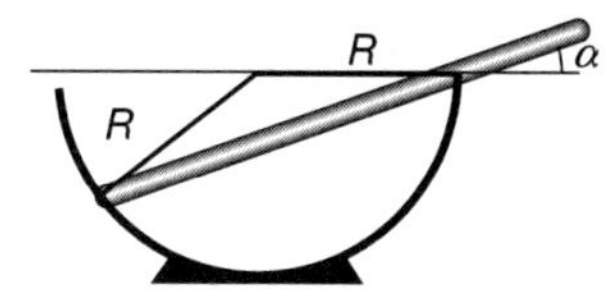

Sample Problem 6.4

If a rod of length $3R$ is placed in a hemispherical bowl of radius R, determine the angle α the rod will make with the horizontal for the rod to be in equilibrium. Neglect friction between the rod and the bowl, and assume that the bowl does not rock. (See figure at right.)

A plot of equation (SP6.4.6), in radians, is found by first defining the equation as a function in an m-file. Create an m-file (select **New** and **M-File** from your menu) using this same name, with the first line indicating that the equation is a function of x, as follows:

```
function y = sp646(x)
% this file defines the function sp646
y = sin(x).^2 - cos(x).^2 +0.75*cos(x);
```

Give this file the name sp646.m. Note that the squares of sine and cosine must be done element by element, so the dot precedes the operation of squaring. Next use fplot, as defined in Chapter 2, to plot the function between 0 and 7 radians by entering the following into the command window:

```
EDU>fplot('sp646',[0,7])      % this plots the function sp6.4.6 from 0 to 7 rad.
```

This command returns the following plot:

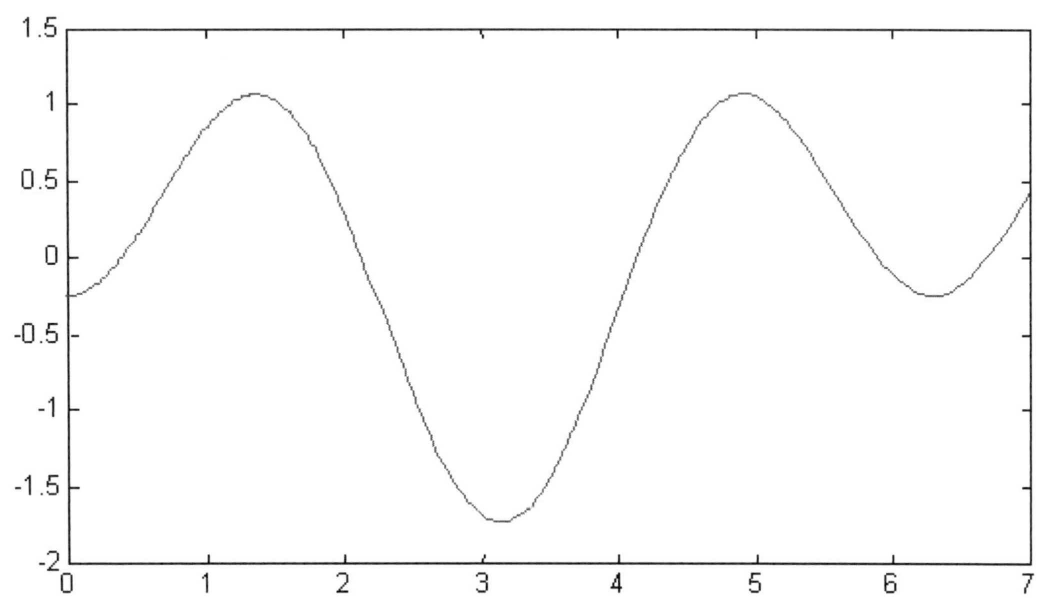

We can see from the plot of sp646 that the lowest root of the equation is approximately 0.4 radian or 23 degrees. This motivates us to plot the function from 0 to 1 rad to confirm our initial guess for use in the fzero command. Note from the plot that if we use 2 radians as our initial guess, the fzero command will converge to the second root of the transcendental equation. If, on the other hand, we use an initial guess of 4 radians, we will determine the third root of the function sp646, and so on. The function is plotted in the following window:

COMPUTATIONAL SOLUTION

```
EDU>fplot('sp646',[0,1])      % plot the function over the interval 0 to 1 rad
EDU>grid                      % turn on the grids so the zero crossing is obvious
```

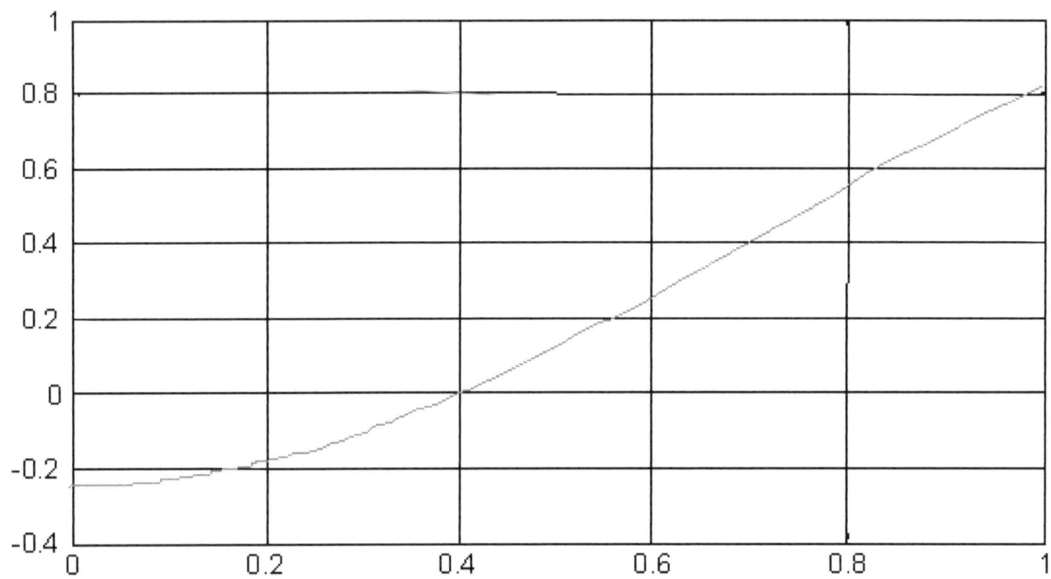


```
EDU>r1=fzero('sp646',0.4)*180/pi   % find the zero near 0.4 rad and change it to
                                   % degrees

r1 =
     23.2133
```

The system of nonlinear equations (SP6.4.1–3) can be solved directly using computational symbolic commands. The symbolic code provided in the *Student Edition of MATLAB* requires that all symbolic quantities be so designated with the syms command and the equations be entered using a label similar to eq#, such that the equations are written with all nonzero terms on the right side. Once entered, the equations are solved by using the solve command. This command returns the solutions stored in matrices for each variable and indicates the number of solutions found. Note that since the equations are nonlinear, there are potentially many solutions. To actually retrieve the form of the solution or its numerical values, additional commands must be entered. In this sample problem, there are three variables (W is set to 1, and the remaining variables are α, N, and P) and four solutions. In the solution vector that results from typing S.P (which lists the values of P symbolically), a numerical value can be obtained by typing double(S.P). This command calls for the numerical value in double precision.

Note that if you are using a different version of MATLAB, the procedures may be slightly different, and you should consult the manual for details of the proper commands to use. The code for computing the solutions of the given function is as follows

COMPUTATIONAL SOLUTION

```
EDU>syms a N P                       % define a, P, and N to be symbolic quantities
EDU>eq1=P*cos(a)-sin(a);             % write out the 3 equations to be solved using
EDU>eq2=N+P*sin(a)-cos(a);           % the eq1-2-3 notation to hold the zero side
EDU>eq3=P*sin(a)*cos(a)*2-cos(a)*(2*cos(a)-1.5);
EDU>S=solve(eq1,eq2,eq3)         % compute the symbolic solution and store it in the
                                 % vector S
S =
    N: [4x1 sym]
    P: [4x1 sym]
    a: [4x1 sym]
EDU>double(S.N)         % evaluate the roots corresponding to N numerically in double
                        precision
ans =
  0.7500
  0.7500
  0.7500
  0.7500
EDU>double(S.P)         % evaluate the roots corresponding to P numerically in double
                        precision
ans =
   1.5423
  -1.5423
   0.4289
  -0.4289
EDU>double(S.a)         % evaluate the roots corresponding to a numerically in double
                        precision
ans =
  2.1460
  2.1460
  0.4051
  0.4051
```

The solution of interest is the one closest to the value $\alpha = 0.3$, rad, which is in the third position corresponding to $P = 0.4298$ and $N = 75$.

Sample Problem 6.5

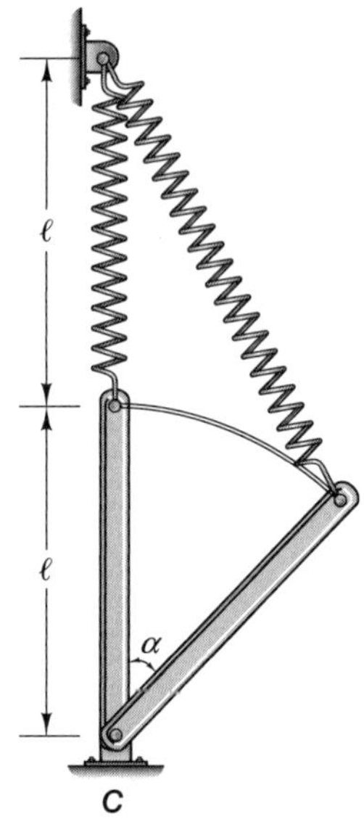

A uniform 2-meter, 50-kg bar is supported by a spring with a spring constant of 500 N/m. The spring is unstretched in the vertical position with a length of 2 meters, as shown in the diagram at the right. Determine the equilibrium angle, α, of the bar.

SOLUTION

A free-body diagram of the bar in the equilibrium position is shown in figure below.

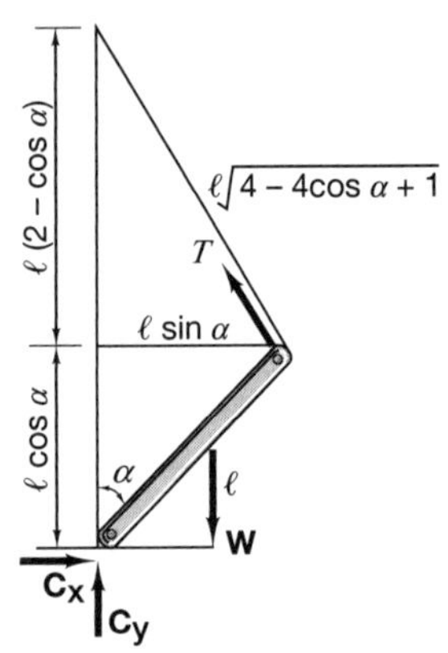

The tension in the spring may be written as:

$$\mathbf{T}(\alpha) = T(\alpha)\left[-\frac{\sin\alpha}{\sqrt{5-4\cos\alpha}}\hat{\mathbf{i}} + \frac{2-\cos\alpha}{\sqrt{5-4\cos\alpha}}\hat{\mathbf{j}}\right]$$

$$\text{where } T(\alpha) = kl\left[\sqrt{5-4\cos\alpha} - 1\right]$$

The weight vector may be written as:

$$W = 50 \cdot g$$
$$\mathbf{W} = -W\hat{\mathbf{j}}$$

The position vectors from point C to $\mathbf{T}$ and $\mathbf{W}$ are:

$$\mathbf{r}_{T/C} = l(\sin\alpha\hat{\mathbf{i}} + \cos\alpha\hat{\mathbf{j}})$$

$$\mathbf{r}_{W/C} = \frac{l}{2}(\sin\alpha\hat{\mathbf{i}} + \cos\alpha\hat{\mathbf{j}})$$

The moment about point C is

$$\mathbf{M_C} = \mathbf{r}_{T/C} \times \mathbf{T} + \mathbf{r}_{W/C} \times \mathbf{W} = \mathbf{0}$$

$$\mathbf{M_C} = \left\{ T(\alpha) l \left[\frac{2 \sin\alpha - \sin\alpha\cos\alpha + \sin\alpha\cos\alpha}{\sqrt{5 - 4\cos\alpha}} \right] - \frac{Wl}{2} \sin\alpha \right\} \hat{\mathbf{k}} = \mathbf{0}$$

Factoring out ($l \sin \alpha$) the solution may be written in the form of the root of a function of α:

$$f(\alpha) = T(\alpha)\frac{2}{\sqrt{5 - 4\cos\alpha}} - \frac{W}{2} = 0$$

This function can be plotted against values of the angle in radians as shown in figure below.

$$\alpha = 0, \frac{\pi}{100} \cdot\cdot \frac{\pi}{5}$$

radians

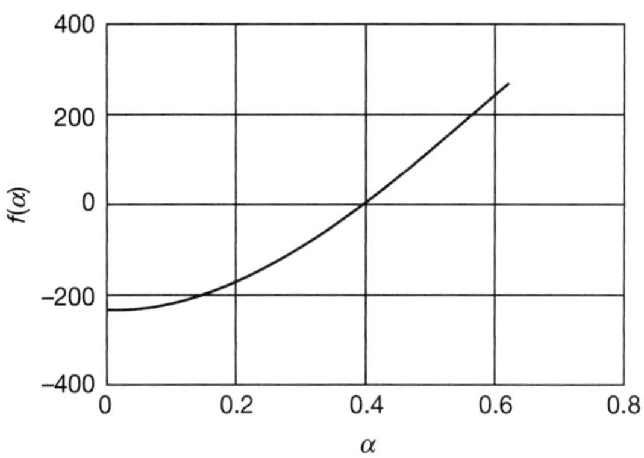

The function is zero at 0.389 radians or 22.3 degrees. Therefore the equilibrium angle is

$$\alpha = 22.3 \text{ degrees}$$

The following typed in the command window yields the solution. Note the analytical solution (`'r='`) for the angle alpha (denoted `'a'` in the code). The solution is in radians.

COMPUTATIONAL SOLUTION: SAMPLE PROBLEM 6.5

```
EDU>W=50*9.81;K=500;L=2;
EDU>r=solve('2*K*L*(sqrt(5-4*cos(x))-1)/(sqrt(5-4*cos(x)))-W/2')
r =
acos(1/4*(64*K^2*L^2-40*K*L*W+5*W^2)/(4*K*L-W)^2)
EDU>a=acos(1/4*(64*K^2*L^2-40*K*L*W+5*W^2)/(4*K*L-W)^2)
a =
    0.3891
```

Sample Problem 6.6

The bar shown below is supported by a ball-and-socket joint at A, by two cables CG and BE, and by a slender rod BF that is attached by means of ball-and-socket joints at both ends. Determine the reactions at the ball-and-socket joint at A, the tension in the two cables,

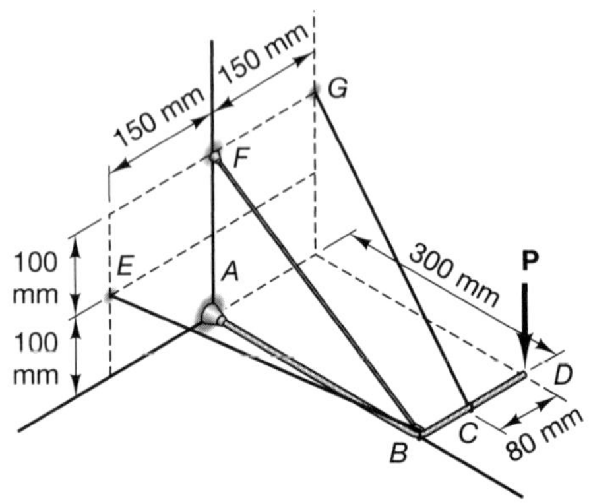

and the force in the rod BF when the system is subjected to a force **P** of magnitude 1000 N. Neglect the weight of the bar.

This problem involves vector algebra and the solution of a system of simultaneous equations. The unit vectors along each of the cables are obtained using the vector algebra capabilities of MATLAB as follows:

```
EDU>rbe=[-300;100;150]; rbf=[-300;200;0]; rcg=[-300;200;-80];  % enter the
                                                                % vectors
EDU>ebe=rbe/norm(rbe),ebf=rbf/norm(rbf), ecg=rcg/norm(rcg)     % compute their
                                                                % unit vectors
ebe =
   -0.8571
    0.2857
    0.4286
ebf =
   -0.8321
    0.5547
         0
ecg =
   -0.8123
    0.5415
   -0.2166
```

The equilibrium equations may be generated using the symbolic processor in MATLAB to compute the force summation and moment summation equations directly from a free-body diagram. The left-hand side of the force-and-moment vector equation is written in symbolic notation and then evaluated symbolically to produce the equations of equilibrium:

SYMBOLIC GENERATION OF EQUILIBRIUM EQUATIONS

```
EDU>syms Ax Ay Az Tbe Tbf Tcg     % declare the unknowns as symbolic variables
                                  % (unit vectors as previously defined)
EDU>SF=[0;-1000;0]+[Ax;Ay;Az]+ebe*Tbe+ebf*Tbf+ecg*Tcg   % enter the forces from
                                                        % the free-body diagram
                                                        % and compute the sum
                                                        % symbolicaly
SF =
[      Ax-857/1000*Tbe-104/125*Tbf-203/250*Tcg]
[ -1000+Ay+67/250*Tbe+111/200*Tbf+271/500*Tcg]
[                 Az+429/1000*Tbe-217/1000*Tcg]
EDU>SM=cross([.3;0;0],(ebe*Tbe+ebf*Tbf))+cross([.3;0;-
  .07],ecg*Tcg)+cross([.3;0;.15],[0;-1000;0])   % enter the moment summation
                                                % from the free-body diagram
                                                % and compute the cross products
                                                % and sums symbolically
SM =
[                          1897/50000*Tcg-150]
[            -1287/10000*Tbe+6097/50000*Tcg]
[ 201/2500*Tbe+333/2000*Tbf+813/5000*Tcg-300]
EDU>pretty(SF),pretty(SM)      % make the equations of equilibrium more readable
     [         857       104        203          ]
     [    Ax - ---- Tbe ---- Tbf - --- Tcg       ]
     [         1000      125        250          ]
     [                                           ]
     [              67         111         271   ]
     [-1000 + Ay + --- Tbe + --- Tbf + --- Tcg]
     [              250        200         500   ]
     [                                           ]
     [              429         217              ]
     [         Az + ---- Tbe - ---- Tcg          ]
     [              1000        1000             ]
       [           1897                        ]
       [           ----- Tcg - 150             ]
       [           50000                       ]
       [                                       ]
```

```
      [           1287          6097             ]
      [       - ----- Tbe + ----- Tcg            ]
      [          10000         50000             ]
      [                                          ]
      [201          333          813             ]
      [---- Tbe + ---- Tbf + ---- Tcg - 300]
      [2500        2000         5000             ]
```

Note that setting each of these expressions equal to zero represents the six equations of equilibrium.

Examining the foregoing symbolic manipulation of the equations of equilibrium enables one to form the coefficient matrix for this system of six linear equations, and the equations can be solved using the matrix inverse, as discussed earlier:

SOLUTION OF THE EQUATIONS OF EQUILIBRIUM

```
EDU>C=[1 0 0 -0.857 -0.832 -0.812; 0 1 0 .286 .555 .542;0 0 1 .492 0 -0.217;0 0
0 0 0 37.907;0 0 0 -128.571 0 121.844; 0 0 0 85.714 166.41 162.429]
                                          % enter the coefficients in C
C =
   1.0000          0          0    -0.8570    -0.8320      -0.8120
        0     1.0000          0     0.2860     0.5550       0.5420
        0          0     1.0000     0.4920          0      -0.2170
        0          0          0          0          0      37.9070
        0          0          0  -128.5710          0     121.8440
        0          0          0    85.7140   166.4100     162.4290
EDU>P=[0;1;0;150;0;300]*1000;
EDU>S=inv(C)*P
S =
  1.0e+03 *
  3.1062
 -0.0021
 -0.7500
  3.7500
 -3.9912
  3.9571
```

Note that the order of this vector is Ax, Ay, Az, T_{bc}, T_{bf}, and T_{cg}, reading from top to bottom. The very first number is a multipier (1.0e+03 *). Thus, the reactions are as follows:

$$\mathbf{A} = Ax\hat{\mathbf{i}} + Ay\hat{\mathbf{j}} + Az\hat{\mathbf{k}} = 3106\hat{\mathbf{i}} - 2.133\hat{\mathbf{j}} - 750.1\hat{\mathbf{k}}$$
$$\mathbf{T}_{be} = 3750(-0.857\hat{\mathbf{i}} + 0.286\hat{\mathbf{j}} + 0.429\hat{\mathbf{k}})$$
$$\mathbf{T}_{bf} = -3991(-0.832\hat{\mathbf{i}} + 0.555\hat{\mathbf{j}})$$
$$\mathbf{T}_{cg} = 3957(-0.812\hat{\mathbf{i}} + 0.542\hat{\mathbf{j}} - 0.217\hat{\mathbf{k}})$$

Note also that T_{bf} is negative so that bar *BF* is in compression.

Sample Problem 6.7

The mechanical system shown in the figure below is used to lift a 100-N weight. A shaft with a radius of 2 m is supported by a thrust bearing at *A* and a nonthrust bearing at *B*. Determine the bearing forces as a function of the angle θ of the crank handle.

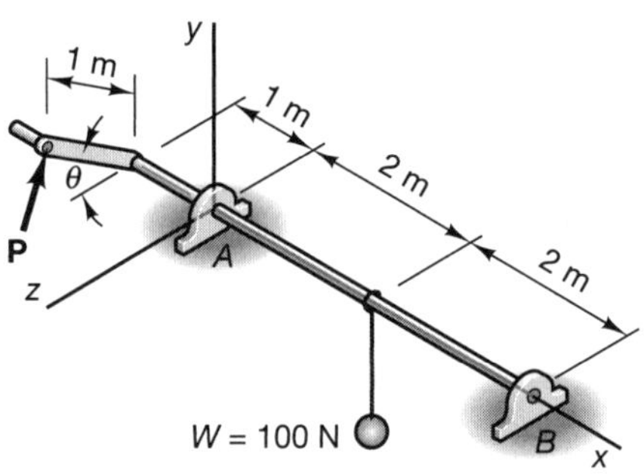

Again, the equilibrium equations can be generated using the symbolic commands:

SYMBOLIC GENERATION OF EQUILIBRIUM EQUATIONS

```
EDU>syms t Ax Ay Az By Bz P  % define the list of symbolic characters
EDU>P*[0;cos(t);-sin(t)]+[Ax;Ay;Az]+[0;-981;0]+[0;By;Bz]  % Compute the
                                                          % resultant force
ans =
[                  Ax]
[ P*cos(t)+Ay-981+By]
[     -P*sin(t)+Az+Bz]
EDU>cross([-1;sin(t);cos(t)],P*[0;cos(t);-sin(t)])+cross([2;0;0.08],[0;981;0])
+cross([4;0;0],[0;By;Bz])          % Compute the sum of the moments symbolically
ans =
```

```
[ -sin(t)^2*P-cos(t)^2*P-1962/25]
[            -P*sin(t)-4*Bz]
[         -P*cos(t)+1962+4*By]
EDU>simple(ans)  % simplify the answer
simplify:
[            -P-1962/25]
[     -P*sin(t)-4*Bz]
[ -P*cos(t)+1962+4*By]
```

Setting the scalar components of the two vector equations equal to zero yields

$$P = 78.5\text{ N} \qquad A_z = 98.13 \sin\theta$$
$$A_x = 0 \qquad B_y = 490.5 + 19.63\cos\theta$$
$$A_y = 490.5 - 98.13\cos\theta \qquad B_z = -19.63\sin\theta$$

The bearing forces can be graphed for one complete revolution of the shaft to determine when they will be a maximum. To set up the plot of each force component and the magnitude of the overall force create an m-file and run it, as indicated in the next window. To do this, simply follow the commands listed in Sample Problem 3.4, which illustrates how to form the magnitude as a function of θ. Note that in this case there are three plots to be put on one graph. Putting these plots on the graph is done by simply writing pairs of values to be plotted together inside the plot statement. (List the value to go along the horizontal first, followed by the function.) A number of plots can be placed on the same graph using this method. To separate the plots, they will each appear in a different color. However, if you will be printing in black and white, you may want to separate the lines by giving them each a different symbol. Immediately following the pair of values to be plotted, enter a comma and set off the symbol code in single quotes, as illustrated in the window that follows. The `xlabel` command is used to label the horizontal axis, and the `title` command is used to indicate what is plotted. Create the following m-file named `bearf`:

COMPUTATIONAL SOLUTION

```
t=(0:10:360)*pi/180;    %sets the range of values for theta, denoted t
A=[490.05-98.13*cos(t);98.13*sin(t)]; % defines vector of bearing forces
mA=sqrt(sum(A.^2)); % computes the magnitude of A as a function of t
plot(t*180/pi,A(1,:),'+',t*180/pi,A(2,:),'*',t*180/pi,mA,'x'),
xlabel('theta in degrees'),
ylabel('forces in Newtons'),
title('Ay denoted +, Az denoted * and |A| denoted x')
```

Then typing `bearf` in the command window results in the following plot:

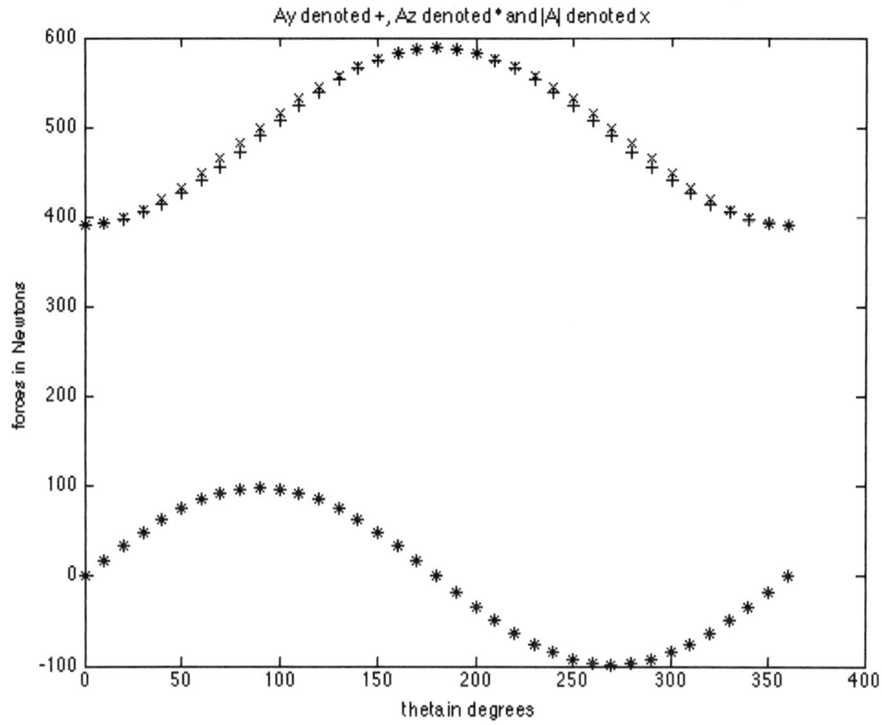

Note that various versions of MATLAB use different plot symbols and color commands.

Sample Problem 6.8

A cam shaft acts against a spring-loaded valve as shown in the diagram at the right. Determine the bearing reactions at A and B and the torsion resisted by the motor for one revolution of the shaft. The bearing at A is a thrust bearing, and the bearing at B does not resist thrust. Model both bearings as simple supports (neglecting bearing moments). The cam is shown in detail just below the shaft. The spring constant $k = 200$ N/m, and the uncompressed length of the spring is to the shaft centerline.

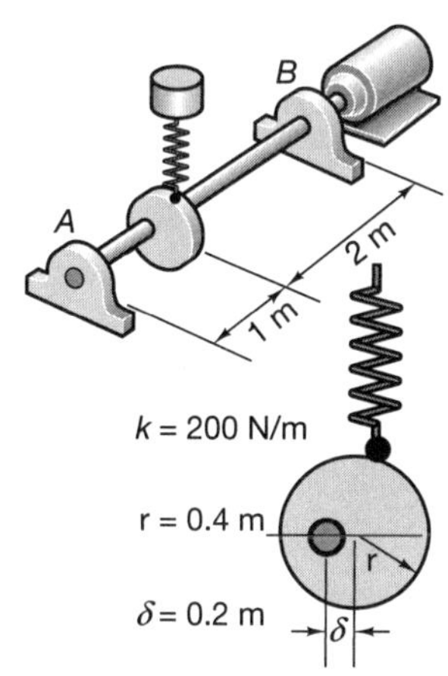

In this problem, we use the symbolic commands to enter equations of equilibrium and solve them symbolically. The cross products are computed symbolically, and you only need to write the force as a vector and the position vector of application; the `cross` and + commands will take the moments and sum them symbolically.

COMPUTATIONAL GENERATION OF EQUILIBRIUM EQUATIONS

```
EDU>syms k ys Ax Ay Az Bx By xs T     % enter the symbolic variables separated
                                      % by spaces
EDU> [0;-k*ys;0]+[Ax;Ay;Az]+[Bx;By;0]    % create the sum of the forces
ans =
[        Ax+Bx]
[ -k*ys+Ay+By]
[           Az]
EDU>cross([xs;ys;0],[0;-k*ys;0])+cross([0;0;1],[Ax;Ay;Az])+cross([0;0;-2],
[Bx;By;0]) +[0;0;T]             % form the sum of the moments and evaluate them
ans =
[   -Ay+2*By]
[    Ax-2*Bx]
[ -xs*k*ys+T]
```

Next, use these results to form a system of linear equations to solve by pasting each of the force and moment expressions into a symbolic equation as follows:

```
EDU>eq1=Ax+Bx;
EDU>eq2= -k*ys+Ay+By;
EDU>eq3=Az;
EDU>eq4=-Ay+2*By;
EDU>eq5=Ax-2*Bx;
EDU>eq6= -xs*k*ys+T;
EDU>S=solve(eq1,eq2,eq3,eq4,eq5,eq6,Ax,Ay,Az,Bx,By,T); % symbolically solve for
  % the variables listed and store the solution in S:
EDU>S.Ax              % display the value for Ax
ans =
0
EDU>S.Ay              % display the value for Ay
ans =
2/3*k*ys
EDU>S.Az              % display the value for Az
ans =
Az
EDU>S.Bx              % display the value for Bx
```

```
ans =
0
EDU>S.By           % display the value for By
ans =
1/3*k*ys
EDU>S.T            % display the value for T
ans =
xs*k*ys
```

The nonzero force components can be plotted for one full revolution of the cam of 360 degrees. This is done by creating the following code in an m-file and then executing it:

```
t=(0:5:360)*pi/180;  % define theta in increments of 5 deg and change to radians
d=.2;r=.4;k=200;     % define the constants d, r, and k
xs=d*cos(t);         % define the numerical function xs
ys=r+d*sin(t);       % define the numerical function ys
Ay=2*k*ys/3;         % define the numerical function Ay
By=k*ys/3; % define the numerical function By
T=k*xs.*ys;% define the numerical function T
plot(t*180/pi,Ay,'x',t*180/pi,By,'*',t*180/pi,T,'+')  % plot and label the
                                                      % functions vs. angel
                                                      % theta converted back to
                                                      % degrees
xlabel('theta in degrees')                            % label the x-axis
title('Ay denote x , By denoted * , and T denoted +') % add a title
```

Saving this m-file and executing it yields the following plot:

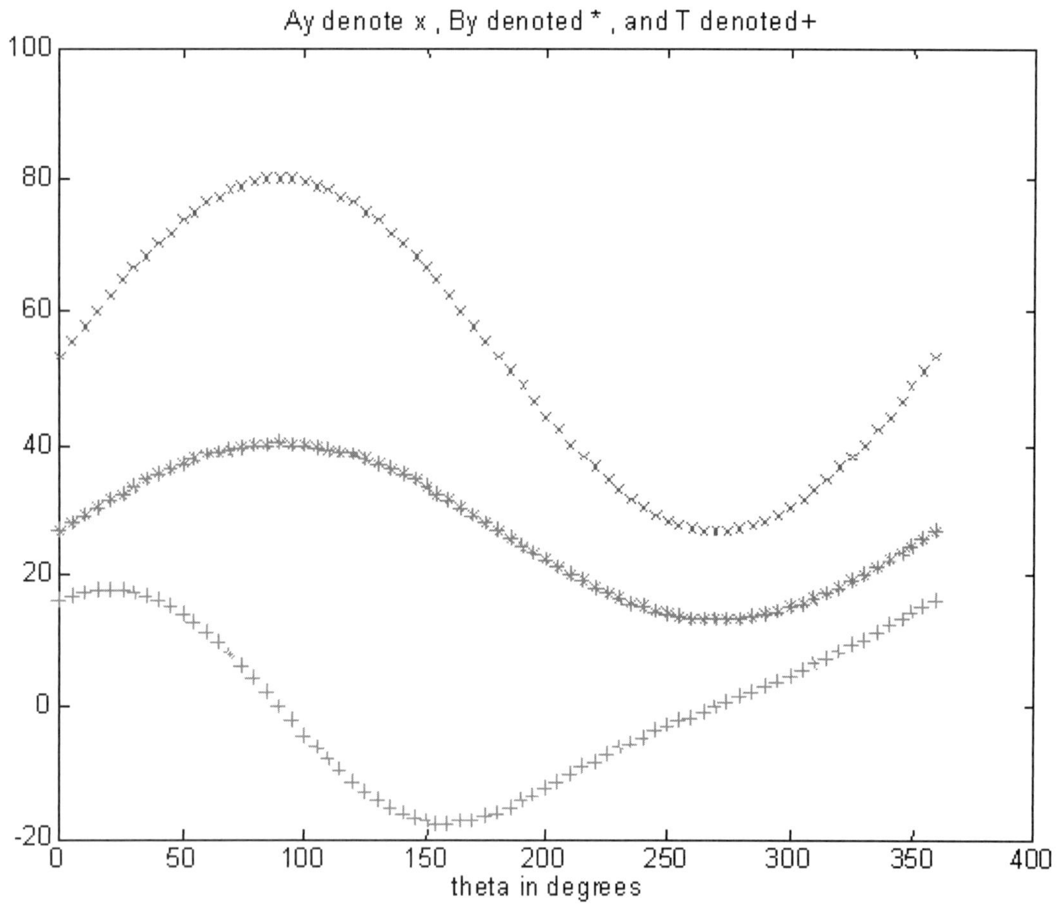

Sample Problem 6.10

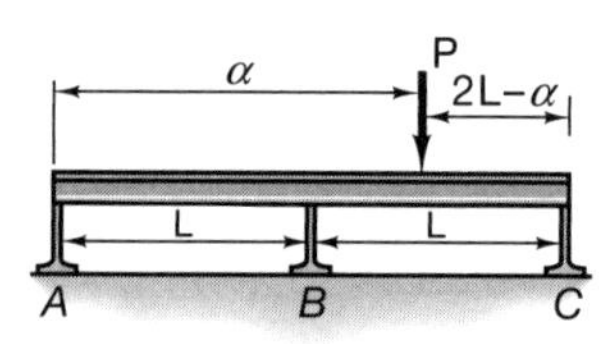

Determine the reactions at the supports A, B, and C for the rigid beam loaded as shown in the diagram at the right.

The three simultaneous equations in Sample Problem 6.10 may be solved symbolically using MATLAB. The approach here is an alternative to that used to solve Sample Problem 6.7, where we used the symbolic commands to solve the equations directly. With Sample Problem 6.10, we write the equations in matrix form and use the symbolic commands to invert the matrix and compute the solution to the simultaneous equations by means of the linear algebra formulation we used numerically before in Section 2:

SYMBOLIC SOLUTION: SAMPLE PROBLEM 6.10

```
syms L P a                     % declare these as symbolic variables
A=[-1 2 -1;1 1 1;0 L 2*L];     % define the symbolic matrix of coefficients
p=[0;P;a*P];                   % define the symbolic vector of loads
inv(A)*p                       % compute the solution by using the symbolic
                               % matrix inverse
```

Executing the foregoing m-file produces the following output:

```
ans =
[  5/6*P-1/2/L*a*P]
[             1/3*P]
[ -1/6*P+1/2/L*a*P]
EDU>pretty(ans)                % make the answer look prettier
                               [              a P ]
                               [ 5/6 P - 1/2 --- ]
                               [              L   ]
                               [                  ]
                               [      1/3 P       ]
                               [                  ]
                               [              a P]
                               [- 1/6 P + 1/2 ---]
                               [              L   ]
```

Sample Problem 6.11

A table has a 30-kg package placed on it, as shown in the figure at the right. If the table has a mass of 50 kg, determine the force in each table leg.

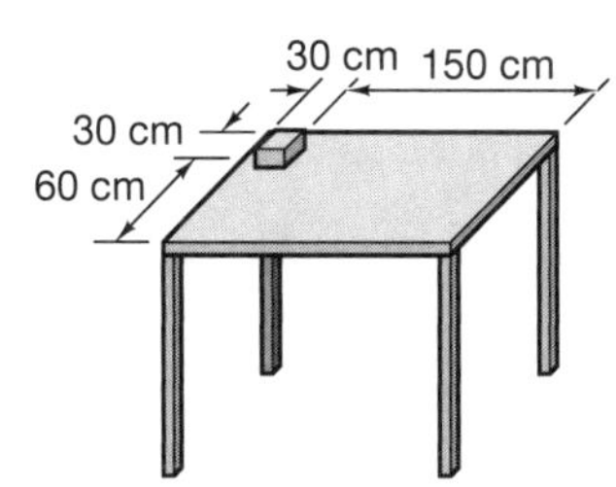

Matrix methods are used to solve the system of six equations in six unknowns of Sample Problem 6.11. In this case, we employ a nonsymbolic numerical approach by using the solution to the equation $Ax = b$, as per Section 2.

COMPUTATIONAL SOLUTION: SAMPLE PROBLEM 6.11

```
>>A=[1 1 1 1 0 0;0 180 180 0 0 0;0 0 -90 -90 0 0;1 -1 0 0 0 180;1 0 -1 0 90 180;
1 0 0 -1 90 0]
A =
     1     1     1     1     0     0
     0   180   180     0     0     0
     0     0   -90   -90     0     0
     1    -1     0     0     0   180
     1     0    -1     0    90   180
     1     0     0    -1    90     0
>>b=[80*9.81;9.81*(30^2+50*90);  -9.81*(30^2+45*50); 0; 0; 0];
%enter the load vector
>>x=inv(A)*b  % compute the solution using the matrix inverse (or use A\b)
x =
   269.7750
   171.6750
   122.6250
   220.7250
    -0.5450
    -0.5450
```

To read the solution from the list of components of the vector x, you must remember the order in which the equations are written. Here x contains the elements (listed in order) A, B, C, D, α, and β. Thus, $A = 269.77$, $B = 171.675$, etc. Note that α and β are given in radians.

Sample Problem 6.12

Consider a bar supported by ball-and-socket joints at A and B, as shown at the right. The bar is loaded by a mass of 100 kg. Determine the support forces at A and B and the tension in the cable CD.

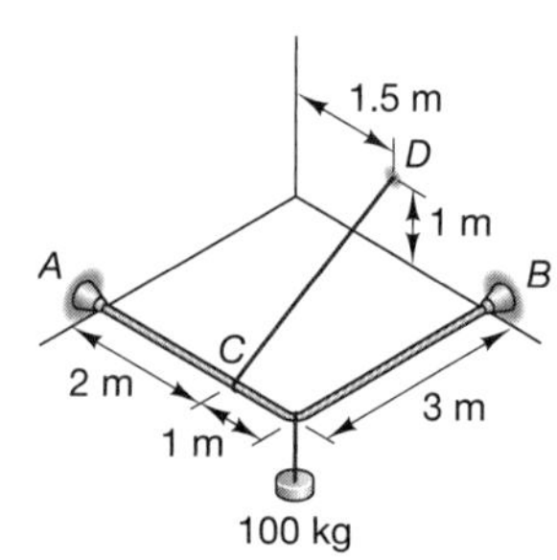

Here we again use the symbolic commands to enter and compute the equations of equilibrium directly from looking at the free-body diagram. This approach follows Sample Problems 6.5 and 6.6.

SYMBOLIC GENERATION OF EQUILIBRIUM EQUATIONS: SAMPLE PROBLEM 6.12

```
EDU>syms Ax Ay Az Bx By Bz T        % name the symbolic variables and form the
                                    % force vectors
EDU>A=[Ax;Ay;Az];B=[Bx;By;Bz];Tv=T*[-0.156;0.312;-0.937];L=[0;-981;0];
EDU>R=A+B+Tv+L;                     % form the vector sum
EDU>M=cross([0;0;3],A)+cross([3;0;0],B)+cross([2;0;3],Tv)+cross([3;0;3],L);
% moments
EDU>M,R                             % display the resultant force and moment
M =
[ -3*Ay-117/125*T+2943]
[  3*Ax-3*Bz+703/500*T]
[    3*By+78/125*T-2943]
R =
[      Ax+Bx-39/250*T]
[ Ay+By+39/125*T-981]
[    Az+Bz-937/1000*T]
```

As discussed in the text, this is a system of six equations in seven unknowns, and the problem, as modeled, is improperly supported. Redesigning the structure by replacing the socket with a cup yields a system of six equations in six unknowns, which is properly supported. The equations of equilibrium can now be symbolically computed from the new free-body diagram as follows:

SYMBOLIC GENERATION OF EQUILIBRIUM EQUATIONS

```
EDU>syms An Ap Ay Bp By T               % enter each of the variables as symbolic
EDU>A=[.707*An+.707*Ap;Ay;-.707*An+.707*Ap];B=[.707*Bp;By;.707*Bp];Tv=
  T*[-0.156;0.312;-0.937];L=[0;-981;0];% enter and define each vector acting on
                                        % the system
EDU>R=A+B+Tv+L:                         % compute the resultant vector
EDU>M=cross([0;0;3],A)+cross([3;0;0],B)+cross([2;0;3],Tv)+cross([3;0;3],L);
                                        % compute the resultant moment
EDU>pretty(R),pretty(M)         % display the sum of forces and moments neatly
        [ 707        707         707        39     ]
        [ ---- An + ---- Ap + ---- Bp - --- T  ]
        [ 1000       1000        1000       250    ]
        [                                          ]
        [                   39                     ]
        [         Ay + By + --- T - 981            ]
        [                   125                    ]
```

```
[                                                  ]
[  707        707        707        937    ]
[- ---- An + ---- Ap + ---- Bp - ---- T]
[  1000       1000       1000       1000   ]

 [              117                   ]
 [       -3 Ay - --- T + 2943         ]
 [              125                   ]
 [                                    ]
 [2121      2121      2121      703   ]
 [---- An + ---- Ap - ---- Bp + --- T]
 [1000      1000      1000      500   ]
 [                                    ]
 [              78                    ]
 [         3 By + --- T - 2943        ]
 [              125                   ]
```

The linear system of equations is solved by writing the six equations in matrix notation and using matrix operations:

```
EDU>C=[.707 0 .707 0 .707 -.156;0 1 0 1 0 .312; -.707 0 .707 0 .707 -.937;0 -3 0
  0 0 -.936;2.121 0 2.121 0 -2.121 1.406;0 0 0 3 0 .624]   % enter the matrix of
                                                           % coefficients
C =
    0.7070         0    0.7070         0    0.7070   -0.1560
         0    1.0000         0    1.0000         0    0.3120
   -0.7070         0    0.7070         0    0.7070   -0.9370
         0   -3.0000         0         0         0   -0.9360
    2.1210         0    2.1210         0   -2.1210    1.4060
         0         0         0    3.0000         0    0.6240
EDU>W=[0;981;0;-2943;0;2943];% enter the loading vector
EDU>inv(C)*W  % compute the solution using the matrix inverse. Note that the
              % answers appear in the following order: An, Ay, Ap, By, Bp, and T.
ans =
   1.0e+03 *
  -2.6050
  -0.4905
   1.5621
   0.0000
   2.0836
   4.7163
```

The tension in the cable could have been determined in both the statically determinate and the statically indeterminate cases by setting the component of the moment about the line AB equal to zero. The reactions at A and B do not produce moments about the line from A to B. This solution is used sometimes, but is based upon a special observation and is not the general solution. The moment about line AB must be zero for equilibrium, and the tension is computed as follows:

COMPUTATIONAL SOLUTION FOR THE TENSION IN THE CABLE

```
EDU>AB=[3;0;-3]; eAB=AB/norm(AB);     % form the unit vector along AB
EDU>syms T                            % declare T as a symbolic variable
EDU>dot(cross([2;0;0],T*[-.156;.312;-.937])+cross([3;0;0],[0;-981;0]),eAB)
    % form the moment of T and W about point A and taking the scalar product
    % with the unit vector along AB yields an expression for the tension T
ans =
-1/2*(-2943+78/125*conj(T))*2^(1/2)
EDU>solve(ans)              % use symbolic solve for the value of T for this
                            % component to be zero
ans =
122625/26
EDU>double(ans)             % change the symbolic solution for T into a
                            % double-precision number
ans =
   4.7163e+03
```

The tension in the cable is the same as previously determined.

7

Analysis of Structures

Application of the equilibrium principles to structures. The structures we use include plane and space trusses, frames, and machines. Most modern texts in structural analysis present methods using the matrix approach. This simplest example of this approach is discussed in Section 7.5 of *Statics*. A symbolic solution of the plane three-member truss in Figure 7.18 is shown in Mathematics Window 7.1 of that text. This symbolic solution can be generated using the symbolic operations in MATLAB.

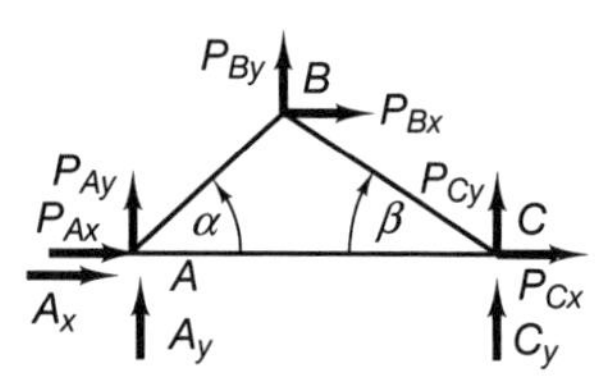

Figure 7.18

The following m-file enters the matrix characterizing the three-member truss for any angles α (a) and β (b), as well as the load vector, all in terms of symbolic variables; then the matrix is inverted and multiplied by the negative of the load vector to obtain symbolic expressions for the internal forces.

```
syms a b Pax Pay Pbx Pby Pcx Pcy
A=[cos(a) 1 0 1 0 0;sin(a) 0 0 0 1 0;-cos(a) 0 cos(b) 0 0 0;
-sin(a) 0 -sin(b) 0 0 0;0 -1 -cos(b) 0 0 0;0 0 sin(b) 0 0 1];
L=[-Pax;-Pay;-Pbx;-Pby;-Pcx;-Pcy];
R=inv(A)*L
```

Executing this file yields the following results

```
R =
[
sin(b)/(cos(a)*sin(b)+sin(a)*cos(b))*Pbx+cos(b)/(cos(a)*sin(b)+sin(a)*cos(b))
*Pby]
[  sin(a)*cos(b)/(cos(a)*sin(b)+sin(a)*cos(b))*Pbx
      -cos(a)*cos(b)/(cos(a)*sin(b)+sin(a)*cos(b))*Pby+Pcx]
[
-
sin(a)/(cos(a)*sin(b)+sin(a)*cos(b))*Pbx+cos(a)/(cos(a)*sin(b)+sin(a)*cos(b))
*Pby]
[                                                             -Pax-Pbx-
Pcx]
[                  -Pay-sin(a)*sin(b)/(cos(a)*sin(b)+sin(a)*cos(b))*Pbx-
sin(a)*cos(b)/(cos(a)*sin(b)+sin(a)*cos(b))*Pby]
[                        sin(a)*sin(b)/(cos(a)*sin(b)+sin(a)*cos(b))*Pbx-
cos(a)*sin(b)/(cos(a)*sin(b)+sin(a)*cos(b))*Pby-Pcy]
```

The truss shown in Figure 7.18 is examined for particular loadings in Computational Windows 7.2 and 7.3. In the first case, a unit vertical load P_{By} is applied, and the internal forces in AB and AC and the vertical reaction A_y as a function of the angle α are determined when $\alpha = \beta$. In the second case, the load in applied at B at an angle of 30 degrees with the vertical, and the internal forces in AB, AC, and BC are examined as a function of the angle α when $\alpha = \beta$.

First consider the case when the triangle is symmetric, so that the two angles are equal, and vary the value of the angle from 5 degrees to 85 degrees. Note that the truss is unstable at 0 degrees and 90 degrees. This is shown in Computational Window 7.2, in which the m-file prepares the functions and plots them when it is called.

COMPUTATIONAL WINDOW 7.2

```
% Create the following m-file named wind7pt2.m:
a=(5:5:85)*pi/180;      % sets the range of values of alpha, denoted by a, from 5
                        % to 85 in degrees and changes it to radians
AB=cos(a)./(2*sin(a).*cos(a));                % defines the functional form for AB
AC=-cos(a).*cos(a)./(2*sin(a).*cos(a));       % defines the functional form for AC
Ay=-sin(a).*cos(a)./(2*sin(a).*cos(a));       % defines the functional form for Ay
plot(a*180/pi,AB,'+',a*180/pi,AC,'*',a*180/pi,Ay,'-')      % plots the functions
                                                           % vs the angle, changed
                                                           % to degrees
xlabel('alpha in degrees')                                 % labels the x-axis
title('AB denoted +, AC denoted *, and the Ay denoted by -')   % titles the figure
                                                               % indicating the key
```

With this m-file saved, typing `wind7pt2` yields the following plot:

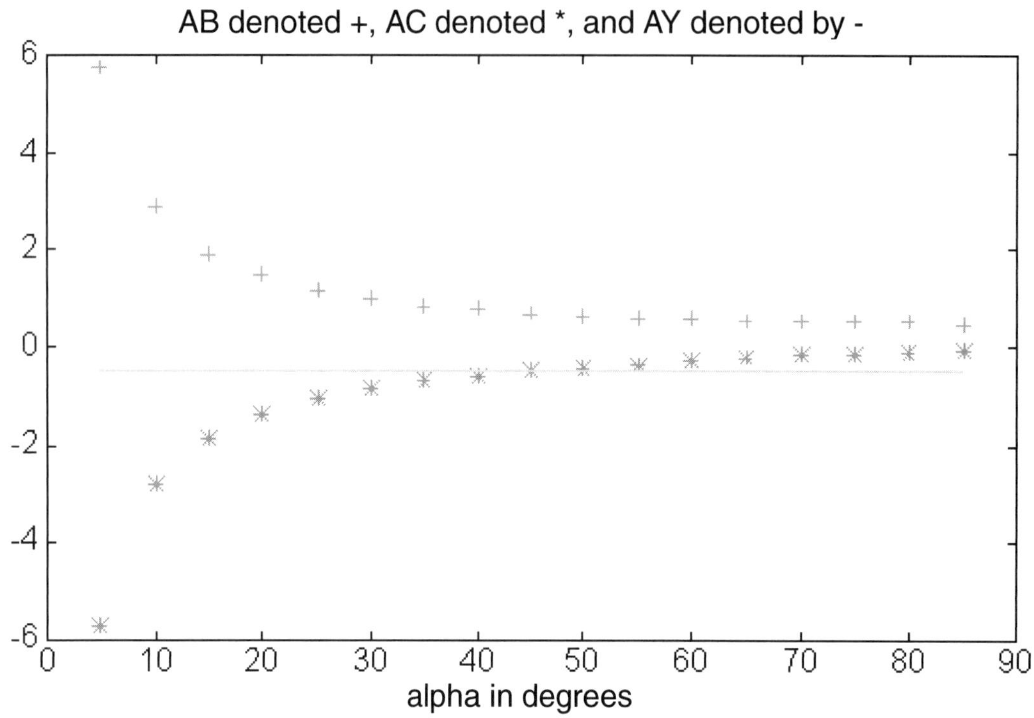

Computational Window 7.3 considers the triangle again to be symmetric with a unit load applied to point *B* at an angle of 30 degrees to the vertical. It also shows the plotted forces in the members as a function of the angle.

COMPUTATIONAL WINDOW 7.3

```
% Create the following m-file named wind7pt3:
a=(5:5:85)*pi/180;      % sets the range of values of alpha, denoted by a, from 5
                        % to 85 in degrees and changes it to radians
PBx=0.5; PBy=0.866;
AB=(sin(a)*PBx+cos(a)*PBy)./(2*sin(a).*cos(a));
AC=(sin(a).*cos(a)*PBx-cos(a).*cos(a)*PBy)./(2*sin(a).*cos(a));
BC=(-sin(a)*PBx+cos(a)*PBy)./(2*sin(a).*cos(a));
plot(a*180/pi,AB,'+',a*180/pi,AC,'*',a*180/pi,BC,'-')
xlabel('alpha in degrees')
title('AB denoted +, AC denoted *, and the BC denoted by -')
```

Typing `wind7pt3` in the command window yields the following plot:

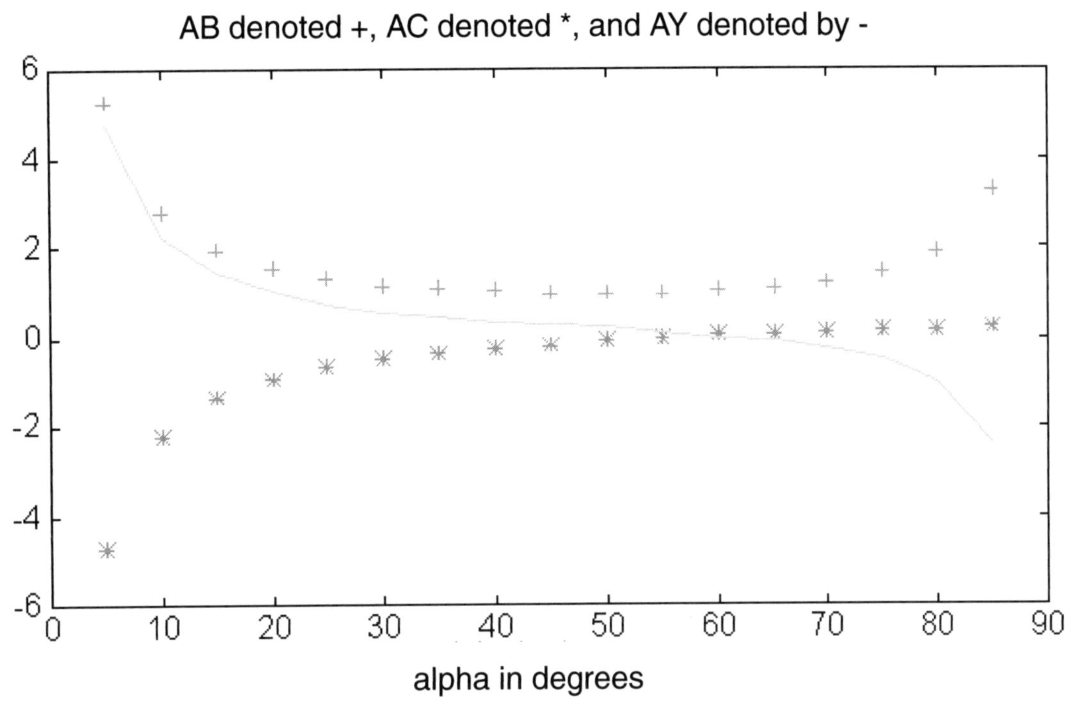

Modern software has rendered obsolete many of the old methods involving techniques that were developed primarily to reduce computational difficulties. For example, the method of sections presented in the *Statics* text is seldom employed anymore in structural analysis. However, MATLAB can still reduce the burden of solving systems of linear equations that arise when the method of sections is used. The system of three equations generated from Figure 7.21 is solved using linear algebra in the plane truss problem.

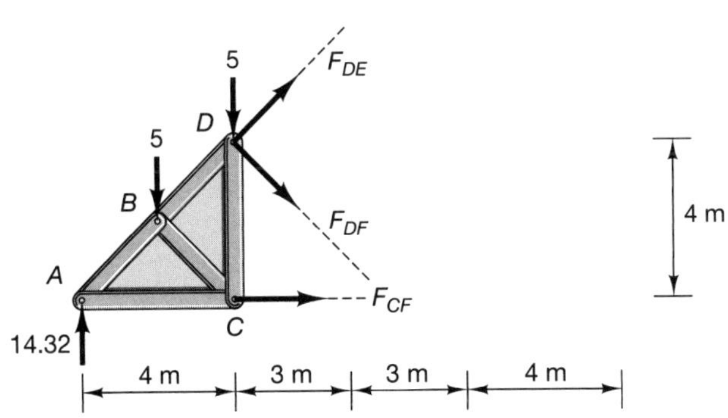

Figure 7.21

COMPUTATIONAL SOLUTION OF THE SECTION SHOWN IN FIGURE 7.21

```
EDU>Ay=14.32; a =45*pi/180;  % enter the constants AY and the angle changing it
                             % radians
EDU>C=[1 3/5 cos(a);0 -4/5 sin(a);4 0 0];   % enter the matrix of coefficients
EDU>C\[0;10-Ay;4*Ay-10]  % solve Cx=b using the more efficient backslash command
ans =
   11.8200
   -5.3571
  -12.1703
```

Note that the order of output of the results is first $CF = 11.82$, followed by $DF = -5.357$ and $DE = -12.17$.

Sample Problem 7.2

Determine the internal member forces and the reactions in the truss shown on the following page, using the method of joints.

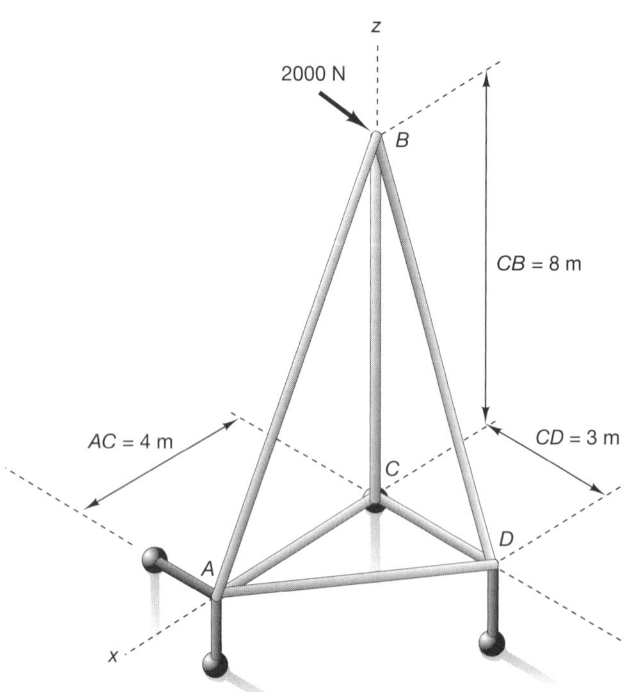

This sample problem requires the solution of a system of 12 linear equations and will be solved using matrix methods set up in an m-file, because of the size of the problem and so that entries can be easily changed to examine various cases. The system of equations is solved for the loading shown in the *Statics* text and for the case when an additional 2000-N load in the x direction is placed at joint B. In the second case, the fourth equation would change. The change can be made by using the $A(i,j)$ referencing system for an array of numbers. This makes it easy to solve a variety of problems and to answer "what if" types of questions.

Enter the following m-file saved as `sp7pt2.m`, which is the matrix of coefficients and the forcing term:

COMPUTATIONAL SOLUTION: SAMPLE PROBLEM 7.2

```
% enter the value of the coefficient matrix and loading function
A=[-.447 -1 -.8 0 0 0 0 0 0 0 0 0; 0 0 .6 0 0 0 1 0 0 0 0 0; .894 0 0 0 0 0 0 1 0
0 0 0; .447 0 0 0 0 0 0 0 0 0 0 0; 0 0 0 0 .352 0 0 0 0 0 0 0;-.894 0 0 -1 -.936
0 0 0 0 0 0 0;0 1 0 0 0 0 0 0 1 0 0 0; 0 0 0 0 1 0 0 0 1 0 0; 0 0 0 1 0 0 0 0 0
0 1 0;0 0 .8 0 0 0 0 0 0 0 0 0;0 0 -.6 0 -.351 -1 0 0 0 0 0 0;0 0 0 0 .936 0 0 0
0 0 0 1];
b=[0;0;0;0;-2000; 0; 0; 0; 0; 0; 0; 0];
X=A\b          % compute the solution to AX=b using matrix left division (LU
               % factorization)
X =
   1.0e+03 *
         0
         0
         0
    5.3182
   -5.6818
    1.9943
         0
         0
         0
   -1.9943
   -5.3182
    5.3182
```

Entering the name of the file or using the execute-and-save function in Mac Operating System (MOS) yields the foregoing list of values. Note that the unknown member forces in the vector **X** are in the order determined by the way in which the equations of equilibrium are written down, which in turn dictates the order of the elements in the matrix **A**. Here the order is

X = [*AB AC AD CB BD CD Ay Az Cx Cy Cz Dz*], so that $CB = 5.318 \times 10^3$, for instance. Next, consider changing the loading and recalculating the solution. Note that in this circumstance, the matrix **A** of coefficients remains the same, and only the fourth element of the loading vector **b** changes, from 0 to 2000. Thus, to calculate the solution for the new loading, open up the m-file `sp7pt2`, and change the fourth element of the vector **b**. Then run the m-file again, as illustrated in the following window:

COMPUTATIONAL SOLUTION: SAMPLE PROBLEM 7.2, CASE 2

```
A=[-.447 -1 -.8 0 0 0 0 0 0 0 0 0; 0 0 .6 0 0 0 1 0 0 0 0 0; .894 0 0 0 0 0 0 1 0
0 0 0; .447 0 0 0 0 0 0 0 0 0 0 0; 0 0 0 0 .352 0 0 0 0 0 0 0;-.894 0 0 -1 -.936
0 0 0 0 0 0 0;0 1 0 0 0 0 0 0 1 0 0 0; 0 0 0 0 0 1 0 0 0 1 0 0; 0 0 0 1 0 0 0 0 0
0 1 0;0 0 .8 0 0 0 0 0 0 0 0 0;0 0 -.6 0 -.351 -1 0 0 0 0 0 0;0 0 0 0 .936 0 0 0
0 0 0 1];   % prepare A from before, as copied
b=[0;0;0;-2000;-2000; 0; 0; 0; 0; 0; 0; 0];    % a new value of b obtained from
                                               % the old, by copying and changing
                                               % the element in the 3rd position
                                               % in b from 0 to 2000.

X =
   1.0e+03 *
   -4.4743
    2.0000
         0
    9.3182
   -5.6818
    1.9943
         0
    4.0000
   -2.0000
   -1.9943
   -9.3182
    5.3182
```

In these last two sample problems, it is important to note that any loading can now be solved without much effort. Once the matrix **A** containing the geometry of the truss is entered, it can be used over and over again to examine the member forces for every possible loading of the system.

Sample Problem 7.2 Using the Direct Vector Approach

Save the following in a file named SP72.m

```
% Enter vectors along members
AB=[-4;0;8];CB=[0;0;8];DB=[0;-3;8];CD=[0;3;0];
AD=[-4;3;0];CA=[4;0;0];
% define unit vectors along each member
cb=CB/norm(CB);db=DB/norm(DB);cd=CD/norm(CD);
ad=AD/norm(AD);ca=CA/norm(CA);ab=AB/norm(AB);
%compute the negative of these:
ba=-ab;bc=-cb;bd=-db;
%compute the biorthogonal vectors
rb=cross(bc,bd);sb=cross(bd,ba);tb=cross(ba,bc);
```

Then enter the following in the command window to compute the forces at pin B:

```
>> SP72
>> PB=[0;2000;0];FAB=-PB'*rb/(ba'*rb),FCB=-
PB'*sb/(bc'*sb),FDB=-PB'*tb/(bd'*tb)

FAB =

     0

FCB =

   5.3333e+03

FDB =

   -5.6960e+03

%Next compute the forces and reaction at D
>> dc=-cd;da=-ad;k=[0;0;1];rd=cross(da,k);
sd=cross(k,dc); td=cross(dc,da);
>> FDC=-rd'*(FDB'*db)/(rd'*dc),
FDA=sd'*(FDB'*db)/(sd'*da),Dz=td'*(FDB'*db)/(td'*k)

FDC =

       2000

FDA =

          0
```

```
Dz =

   -5.3333e+03
%Next compute the forces at pin A
>>ac=[1;0;0];j=[0;1;0];ra=cross(j,k);sa=cross(k,ac);
ta=cross(ac,j);
>> FAC=-ra'*(FDA'*ad+FAB'*ab)/(ra'*ac),
Ay=-sa'*(FDA'*ad+FAB'*ab)/(sa'*j),
Az=-ta'*(FDA'*ad+FAB'*ab)/(ta'*k)

FAC =

     0

Ay =

     0

Az =

     0

%Last compute the reactions at joint C
>> Cx=-FAC,Cy=-FDC,Cz=-FCB

Cx =

     0

Cy =

         -2000

Cz =

   -5.3333e+03
```

Note that the answers in the direct vector solution are slightly different from those obtained using a matrix inverse.

Sample Problem 7.3

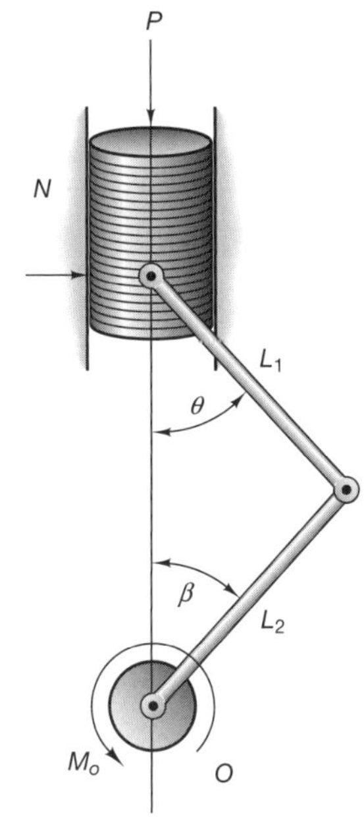

In the figure at the right, variable torque motor is programmed to maintain a constant force P on the piston as the shaft rotates a full revolution. Determine the torque as a function of the shaft rotation angle β. Neglect the inertia of the moving parts, and solve for the quasistatic case. Graph the required torque for equilibrium for $L_2 = 500$ mm, $L_1 = 200$ mm, and $P = 1000$ N.

To fully understand the solution of Sample Problem 7.3, the motor torque should be plotted for a full revolution. Note that L1 must be greater than L2, or the machine will lock up at 90 degrees. While L2 rotates through 360 degrees, L1 oscillates though a fixed arc. In addition, the dependence of the solution on the length 3 L1 and L2 can be investigated by entering these as separate constants in an m-file. Here we will write an m-file to compute the motor torque and then plot it for given values of L1 and L2, as indicated in the following window:

MOTOR TORQUE GRAPH FOR SAMPLE PROBLEM 7.3 (L2 = 200 MM)

```
%this file computes the torque for sample 7.3
L1= 500; L2=200; P=1000;          % enter the values of the constants
b=0:0.1:2*pi;                     % define beta, denoted b, to range from 0 to 2pi
                                  % radians
M=P*(L1*cos(asin(L2*sin(b)/L1))+L2*cos(b)).*tan(asin(L2*sin(b)/L1));
plot(b*180/pi,M)                  % change b to degrees and plot the torque versus
                                  % beta
xlabel('beta in degrees')                     % label the x-axis
title('Motor torque as a function of angle')  % give the plot a title
```

Using `save and execute` or entering the name of this m-file yields the following plot:

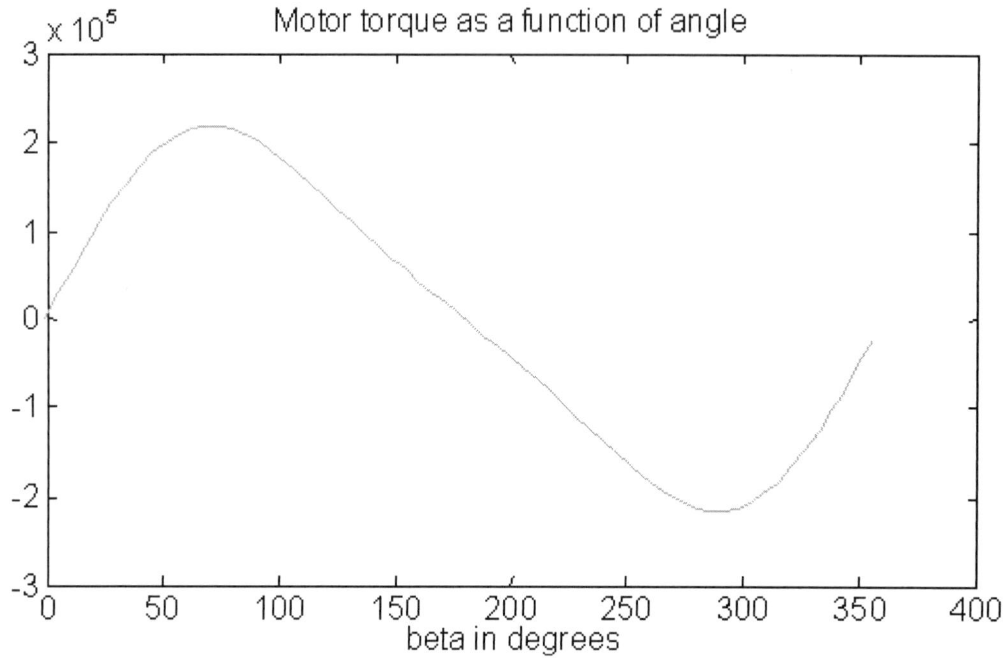

Note that the moment required for equilibrium is zero when the piston is at the top or at the bottom of its stroke and that the moment is negative during the second half of the cycle.

The effect of changing the ratio of L1 to L2 can be easily examined. Once the MATLAB m-file for a problem is created, a change in a constant, such as L2, can be accomplished by opening the file and simply changing the value. Running the file again by typing the file name after the prompt will produce the new plot or solution. Suppose the length of L2 is changed to 300 mm. Then the required moment would increase as shown in the following plot

COMPUTATIONAL SOLUTION : SAMPLE PROBLEM 7.3 (L2 = 300 MM)

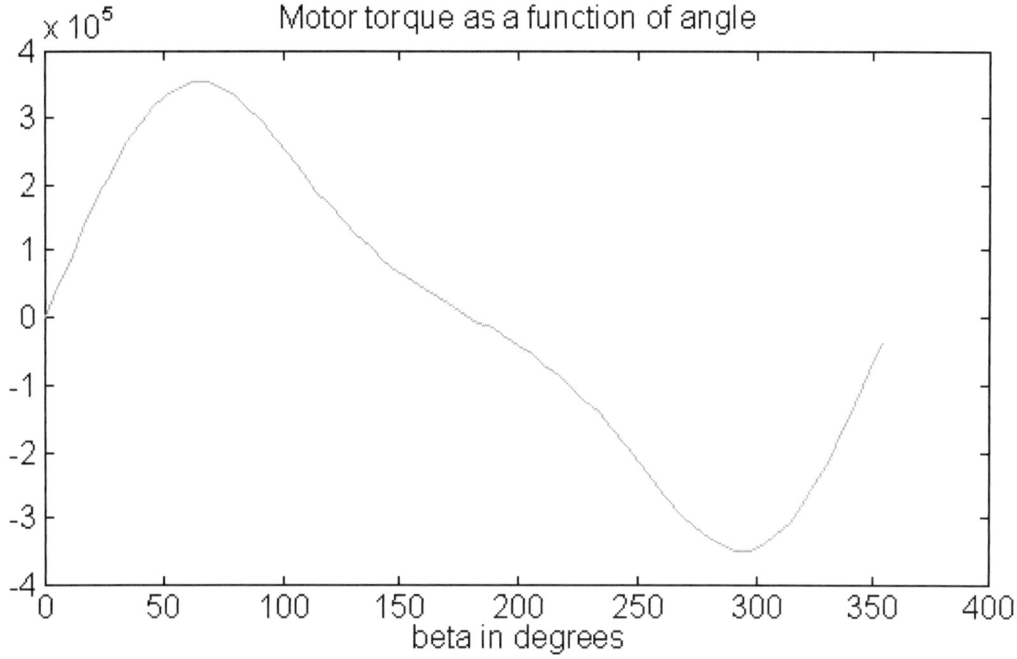

Note that as the length L2 is increased, the moment for equilibrium also increases.

Sample Problem 7.4

A compound-lever cutting tool is shown in the accompanying diagram. A force F is applied to each of the handles. Determine the cutting force between E and G, the pin reactions at C and D, and the tension or compression in member BH. All dimensions are in millimeters.

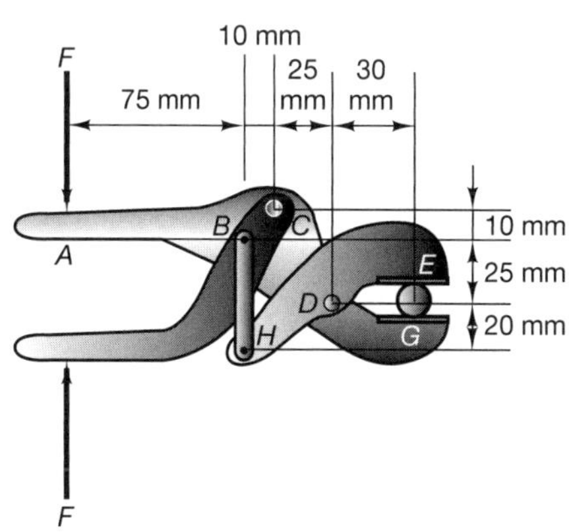

This sample problem is solved using different combinations of the equilibrium equations. First we symbolically solve equations (a) and (b):

COMPUTATIONAL SOLUTION a & b

```
EDU>syms F f A    % define these as symbolic variables, followed by the matrix A
EDU>A=[1 0 1 0 0 0;0 1 0 1 0 -1; 0 0 25 25 0 -55;0 0 -1 0 0 0;0 0 0 -1 1 1;0 0 0
0 -35 30]
A =
     1     0     1     0     0     0
     0     1     0     1     0    -1
     0     0    25    25     0   -55
     0     0    -1     0     0     0
     0     0     0    -1     1     1
     0     0     0     0   -35    30
EDU>f=[0;F;-85*F;0;0;0];        % enter the load vector
EDU>pretty(inv(A)*f)            % compute the solution and make it pretty
                             [   0    ]
                             [        ]
                             [- 15/2 F]
                             [        ]
                             [   0    ]
                             [        ]
                             [ 221    ]
                             [ --- F  ]
                             [ 12     ]
                             [        ]
                             [ 17/2 F ]
                             [        ]
                             [ 119    ]
                             [ --- F  ]
                             [ 12     ]
```

Using equations (a) and (c) yields the following results:

COMPUTATIONAL SOLUTION a & c

```
EDU>syms F
EDU>B=[1 0 1 0 0 0;0 1 0 1 0 -1;0 0 25 25 0 -55;1 0 0 0 0 0;0 -1 0 0 -1 0;0 -10 0
  0 0 0]
B =
     1     0     1     0     0     0
     0     1     0     1     0    -1
     0     0    25    25     0   -55
     1     0     0     0     0     0
     0    -1     0     0    -1     0
     0   -10     0     0     0     0
EDU>f=[0;F;-85*F;0; -F; 75*F];
EDU>pretty(inv(B)*f)
                              [    0     ]
                              [          ]
                              [- 15/2 F]
                              [          ]
                              [    0     ]
                              [          ]
                              [ 221      ]
                              [ --- F    ]
                              [ 12       ]
                              [          ]
                              [ 17/2 F   ]
                              [          ]
                              [ 119      ]
                              [ --- F    ]
                              [ 12       ]
```

We also could have used equations (b) and (c), as illustrated in the next window:

COMPUTATIONAL SOLUTION b & c

```
EDU>B(1,3)=-1;B(2,2)=0;B(2,4)=-1;B(2,5)=1;B(2,6)=1;B(3,4)=0;
  B(3,5)=-35;B(3,6)=30;
EDU>B
B =
     1     0    -1     0     0     0
     0     0     0    -1     1     1
     0     0    25     0   -35    30
     1     0     0     0     0     0
     0    -1     0     0    -1     0
     0   -10     0     0     0     0
EDU>f(1)=0;f(2)=0;f(3)=0;f(4)=0;
EDU>pretty(inv(B)*f)
                         [    0     ]
                         [          ]
                         [- 15/2 F]
                         [          ]
                         [    0     ]
                         [          ]
                         [ 221      ]
                         [ --- F    ]
                         [ 12       ]
                         [          ]
                         [ 17/2 F   ]
                         [          ]
                         [ 119      ]
                         [ --- F    ]
                         [ 12       ]
```

8

Internal Forces in Structural Members

In this chapter, we use the principle of equilibrium to determine and diagram the internal forces in structural members. The chapter presents the basic concepts of beam loading in preparation for the usual follow-on course in strength of materials. Most of the integrands required to determine the internal shears and moments in beams should be integrated numerically or symbolically using MATLAB. The shear and moment diagrams, or plots, can then be generated. Sample Problem 8.4 is solved using both methods.

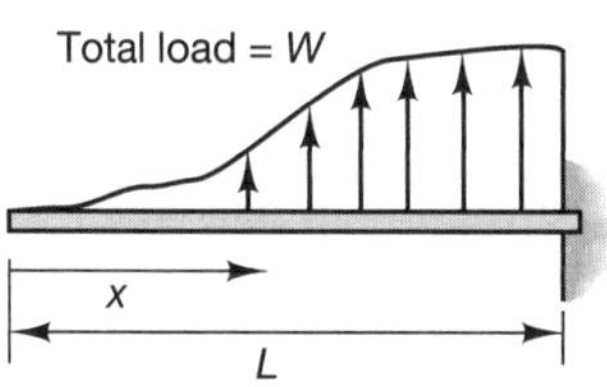

Sample Problem 8.4

Detemine the shear and moment equations for a ***cantilever beam*** loaded by a load function

$$\text{(a) } w(x) = kx \quad \text{where } k = W/L$$
$$\text{(b) } w(x) = W\sin(\pi x/L)$$

where W is the maximum load intensity, given in N/m, and L is the length of the beam in meters. (See diagrams at right.)

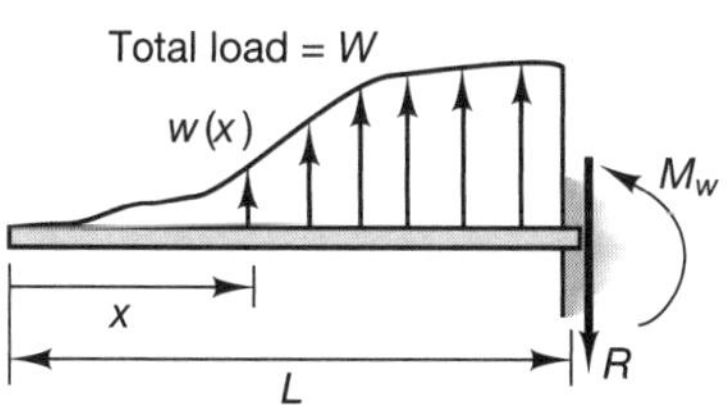

This first window shows the symbolic calculation of the shear and moment:

SAMPLE PROBLEM 8.4: SYMBOLIC SOLUTION

```
EDU>syms W L u x          % define the symbolic variables
EDU>f1=(W/L)*u;f2=(x-u)*u*W/L;f3=W*sin(pi*u/L);f4=(x-u)*W*sin(pi*u/L);
EDU>Va=int(f1,u,0,x); Ma=int(f2,u,0,x);Vb=int(f3,u,0,x);Mb=int(f4,u,0,x);
EDU>pretty(Va)
                          2
                         x  W
                     1/2 ----
                          L
EDU>pretty(Ma)
                          3
                         x  W
                     1/6 ----
                          L
EDU>pretty(Vb)
                     x pi
                 cos(----) L W
                      L         L W
             - ------------- + ---
                     pi         pi
EDU>pretty(Mb)
               2      x pi
              L  sin(----) W
                       L        x L W
            - -------------- + -----
                       2        pi
                     pi
```

The foregoing integrals were evaluated symbolically. Next, we examine the evaluation and plotting of the shear and moment diagram using numerical computations. Again, we will use the quad8 numerical integration scheme, which requires setting up the integrands in m-files. The following two files, named f812.m and f822.m, respectively, contain the required integrands:

```
function y=f812(u,W,L)        % this is the file containing the integrand for V,
                              % the shear, as a function of u,W and L
wu = W/L*u;
y = wu;
```

```
function y=f822(u,x,W,L)       % defines the integrand for M
                               % given by (x-u)*W/L*u as a function of u, x, W
                               % and L.
wu = W/L*u;
y = (x-u).*wu;
```

Now we write the following m-file to call the previous two functions, numerically integrate them, and plot the shear and moment diagrams:

NUMERICAL SOLUTION OF PROBLEM 8.4

```
% this file sets up the integrals for V(x) and M(x) as defined in the m-files
f812.m and f822.m
W = 200; L = 10 ;          % define the constants W and L
x=0:.1:10;                 % define variable upper limit of integration
n=length(x);               % sets the number of integrations and stores it in n
% numerical integration; repeated for each individual value of the upper limit
x(i)
for i = 1:n;               % starts the loop that integrates over the varialbe limit
                           % of integration
   Vx(i) = quad8('f812',0,x(i),[],[],W,L);         % calculates the shear
   Mx(i) = quad8('f822',0,x(i),[],[],x(i),W,L);    % calculates the moment
end;                                               % ends the loop
figure(1); clg;                                    % plot functions in figure 1
subplot(2,1,1)                                     % sets up two different plots,
                                                   % and sets up the first
   plot(x,Vx)                                      % plots V(x) versus x
   xlabel('x')                                     % labels the x-axis
   ylabel('V(x)')                                  % labels the y-axis
subplot(2,1,2)                                     % sets up the second plot
   plot(x,Mx)                                      % plots M(x) versus x
   xlabel('x')                                     % labels the x-axis
   ylabel('M(x)')                                  % labels the y-axis
```

Entering `save and execute` on a Macintosh, or saving under the name `sp8pt4.m` and then entering `sp8pt4` after the prompt on any machine, yields the following plots:

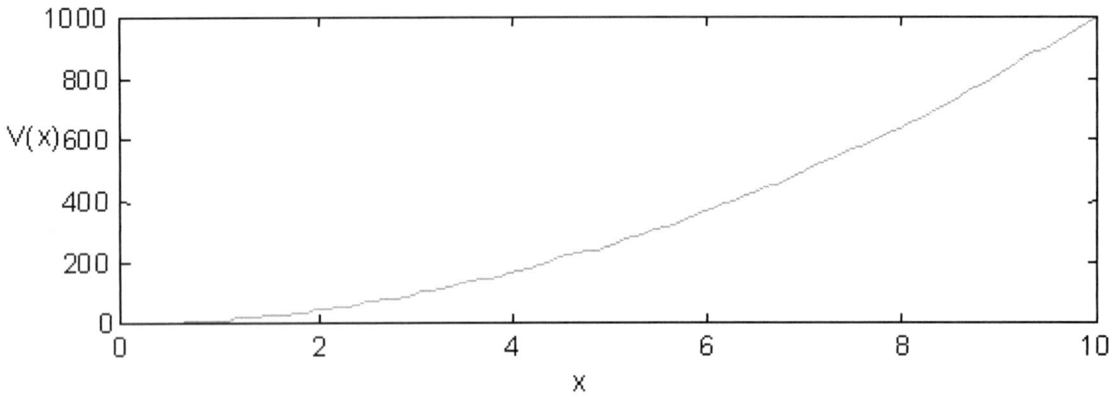

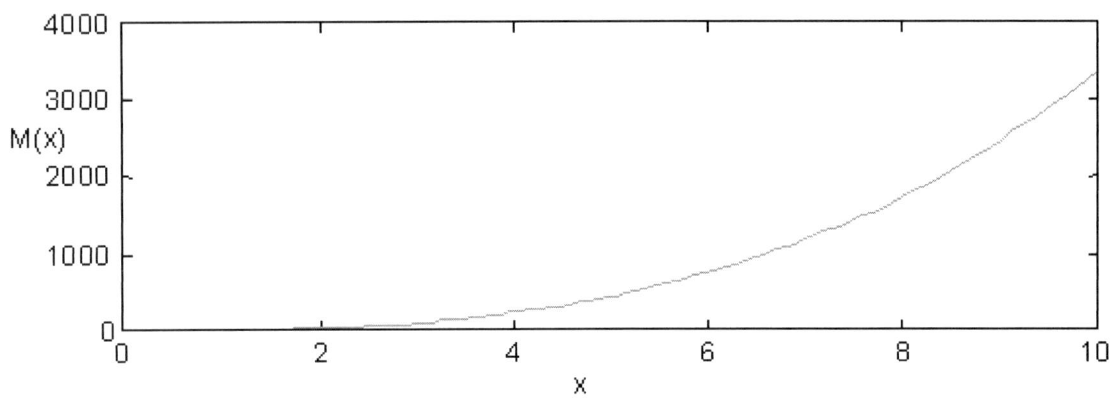

Next, consider situation "b", in which $w(x) = W\sin(\pi x/L)$, and calculate the resulting shear and moment diagram. Again, set up an m-file with the integrand for $V(x)$ and another for the integrand of $M(x)$. The shear m-file is displayed as follows:

```
function y=f8pt4aV(u,W,L)  % the integrand for V(x), case b, Sample Problem 8.4
y=W*sin(pi*u/L);
```

The m-file is stored as f8pt4aV.m. The m-file for the integrand of the moment is the following:

```
function y=f8pt4aM(u,x,W,L) % the integrand for M(x), case b, Sample Problem 8.4
y= W*(x-u).*sin(pi*u/L);
```

This m-file is saved as f8pt4aM.m. Note that the order of the functional dependence is critical.

Now we create an m-file to integrate these functions, which essentially are the same as those in the preceding files, but with the file names changed in the quad8 command. This new m-file, saved as sp8pt2b.m, is as follows:

```
% Numerical solution to problem 8.4a
P = 200; L = 10;
% define variable upper limit of integration
x=0:.5:10;
% number of integrations = n
n=length(x);
% numerical integration; repeated for each individual value of the upper limit x(i)
for i = 1: n;
    Px(i)= P*sin(pi*x(i)/L);
    Vx(i) = quad8('f8pt4aV',0,x(i),[],[],P,L);
    Mx(i) = quad8('f8pt4aM',0,x(i),[],[],x(i),P,L);
end;
% plot functions in figure 1
figure(1); clg;
subplot(3,1,1)
plot(x,Px), xlabel('x'), text(1,100,'P(x)')     % plot the load and label it
subplot(3,1,2)
plot(x,Vx), xlabel('x'), text(1,500,'V(x)')     % plot the shear and label it
subplot(3,1,3)
plot(x,Mx), xlabel('x'),text(1,5000,'M(x)')     % plot the moment and label it
```

Saving and executing this file returns the following plots

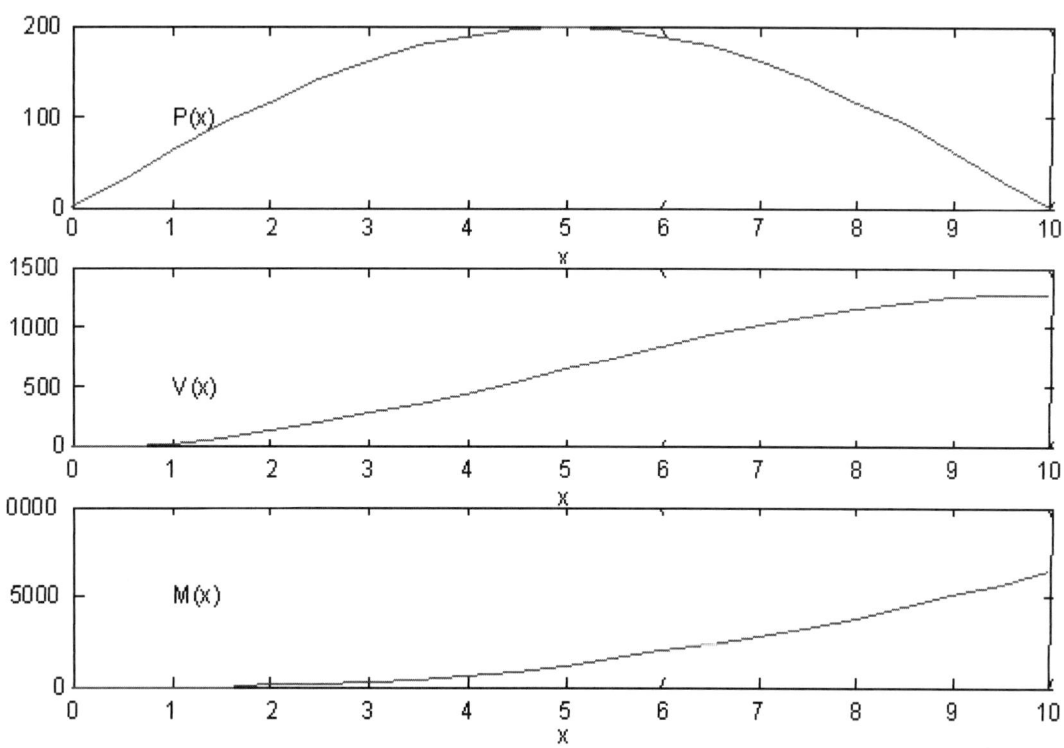

Sample Problem 8.6

Write the shear and moment equations for the beam shown in the diagram at the right.

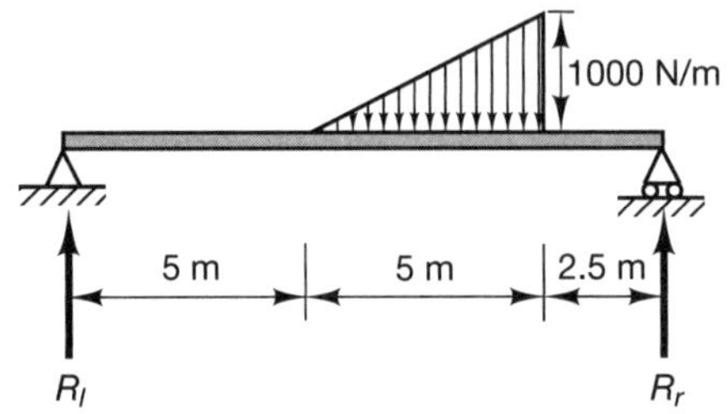

The maximum moment in Sample Problem 8.6 can be determined using the fzero command to solve the quadratic equation for the moment. First, plot the shear as a function of x to obtain an initial guess of the root. The following m-file, saved under the name sp8pt6.m, plots the shear:

```
x=5:0.1:10;
M=833-200*0.5*(x-5).^2;
plot(x,M)
xlabel('x')
ylabel('M(x)')
```

Then, entering sp8pt6 after the prompt produces the following plot:

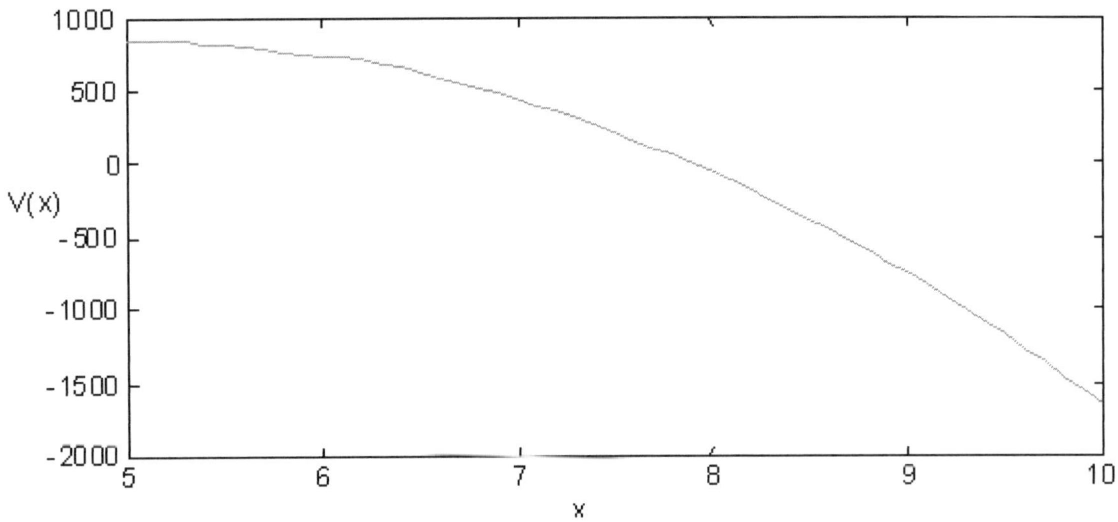

The plot reveals that $V(x)$ passes through zero around $x = 8$, which now forms the initial guess for the fzero command used in the code that follows to compute the exact root. First, set up an m-file with the functional form of $V(x)$ in it as follows:

```
function V=f8pt6(x)
V=833-0.5*200*(x-5)^2;
```

```
EDU>a=fzero("f8pt6',8) % look for a root near 8
a =
    7.8862
```

```
function M=M8pt6(x)
M=833*x-(200/6)*(x-5)^3;
```

With the moment function defined in the previous m-file, entering the following yields the maximum value of the moment at $x = a$:

```
EDU>MaxM=M8pt6(a)
MaxM =
   5.7678e+03
```

Next, define the moment function numerically, and plot it by executing the following m-file:

```
x=5:.1:10;
M=833*x-(200/6)*(x-5).^3;
plot(x,M),grid
xlabel('x')
ylabel('M')
```

Running this file produces the following plot:

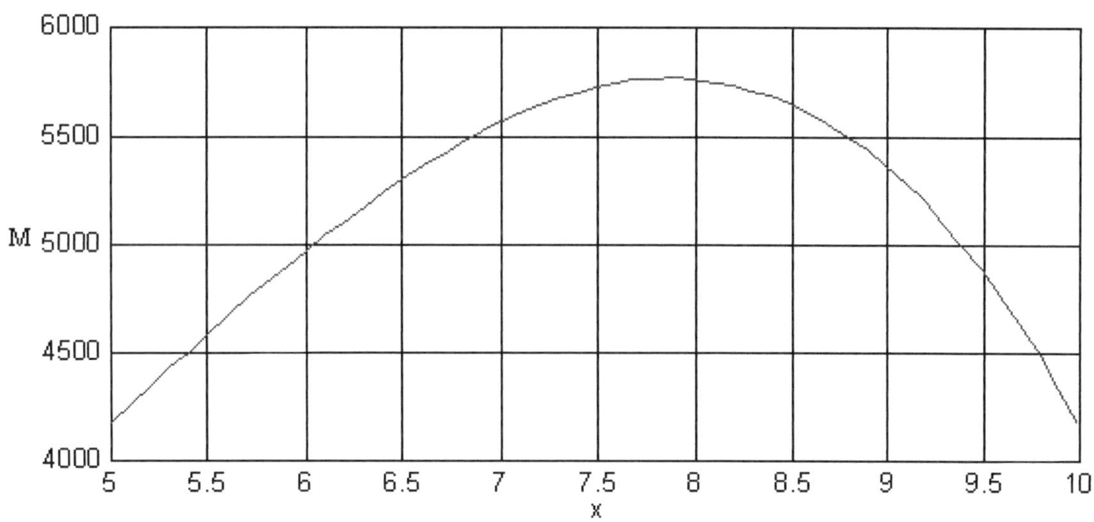

This plot shows clearly that the maximum occurs near the value $x = 7.886$, as previously calculated.

The shear and moment diagrams for Sample Problem 8.6 can be generated using a different range of the variable x for each section of the beam. By using the `'hold on'` command in MATLAB and by defining x, $V(x)$, and $M(x)$ over the three sections of the beam, the composite shear and moment diagrams may be plotted. The `hold on` command in MATLAB allows several curves to be calculated separately, but plotted on the same graph. In this case, each shear function is calculated and plotted over a region of x values, and the hold on function is used to plot that segment and then pause until the next segment is ready to plot, scaling the axis accordingly. The following m-file, named sm8pt6.m, calculates the desired shear and moment and sets them up to be plotted.

```
x2=5:.5:10;
V2=833-.5*200*(x2-5).^2;
M2=833*x2-(200/6)*(x2-5).^3;
subplot(2,1,1)
plot(x2,V2)
hold on
subplot(2,1,2)
plot(x2,M2)
hold on
x3=10:.5:12.5;
V3=-1667*x3./x3;
M3=-1667*x3+20833;
subplot(2,1,1)
plot(x3,V3)
xlabel('x')
ylabel('V(x)')
subplot(2,1,2)
plot(x3,M3)
xlabel('x')
ylabel('M(x)')
```

Typing sm8pt6 after the prompt produces the following plots, historically known as the shear and moment diagrams:

SHEAR AND MOMENT DIAGRAMS: SAMPLE PROBLEM 8.6

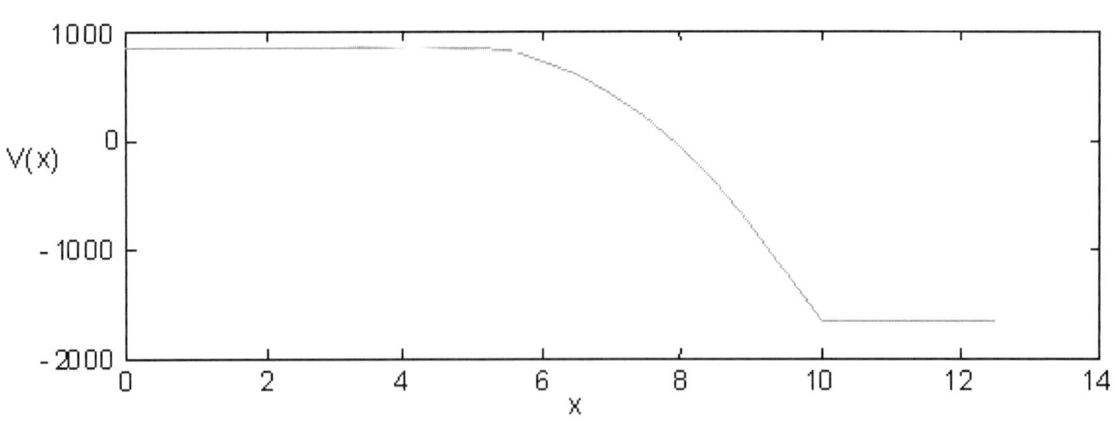

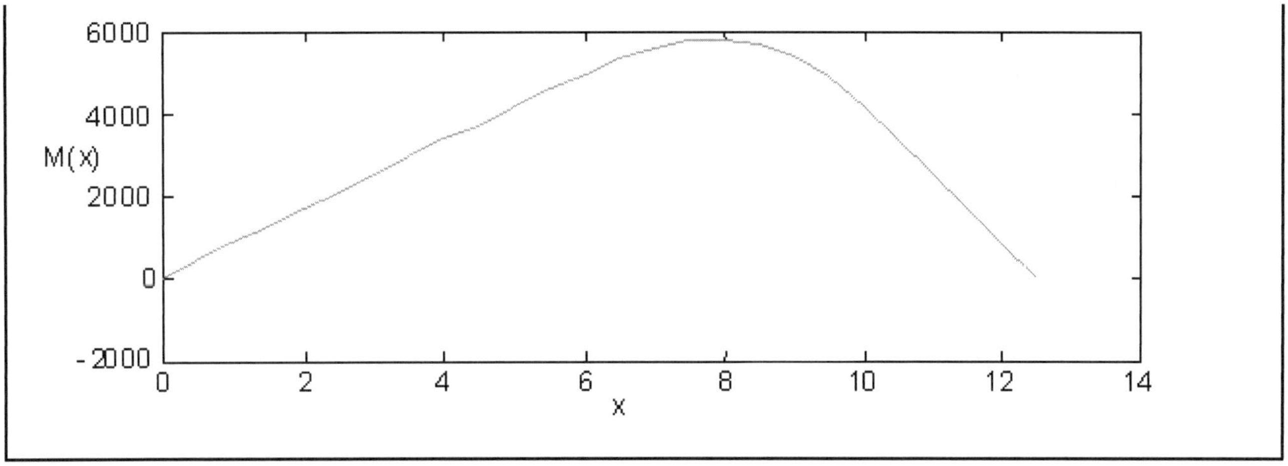

Discontinuity Functions

MATLAB has the Heaviside step function, designated by `Heaviside (x-a)`, where `a` is a constant and `x` is a variable. If the argument of this function is negative, MATLAB sets the function to zero. If the argument is positive, the Heaviside function is equal to unity. Other discontinuity functions can be generated using the Heaviside function. However, the Dirac delta function is directly available as the command `Dirac(x-a)`.

Sample Problem 8.7

The code that follows uses the `Heaviside` command to define the shear $V(x)$ and plot it showing its dependence on x. In version 5 of MATLAB, the Heaviside command is defined only symbolically, so you must first create an m-file that defines a numerical version of the Heaviside function. This is easily done, since the singularity step function is already embodied in the `stepfun` command. Accordingly, define the following function file and save it as `Heaviside.m`:

```
function y = Heaviside(t)
% uses stepfun to define the heaviside function numerically
y=stepfun(t,0);
```

Next, use the symbolic commands to enter the shear and moment functions using the symbolic `Heaviside` command, as illustrated in the following m-file:

```
syms x    % define the variable x to be symbolic, followed by the shear function
V=833*sym(‘Heaviside(x)‘)-100*sym(‘Heaviside(x-5)‘)*(x-
5)^2+1000*sym(‘Heaviside(x-10)‘)*(x-10)+100*sym(‘Heaviside(x-10)‘)*(x-10)^2;
M=int(V);   % use symbolic integration to compute the moment symbolically
subplot(2,1,1)                  % set up two plots
ezplot(V,[0,12.5])              % plot the shear
title(‘Shear versus x‘)         % add a title
xlabel(‘x‘)                     % label the x-axis
subplot(2,1,2)                  % set up the second plot
ezplot(M,[0,12.5])              % plot the moment
title(‘Moment versus x‘)        % add the title
```

SHEAR AND MOMENT DIAGRAM FOR SAMPLE PROBLEM 8.7

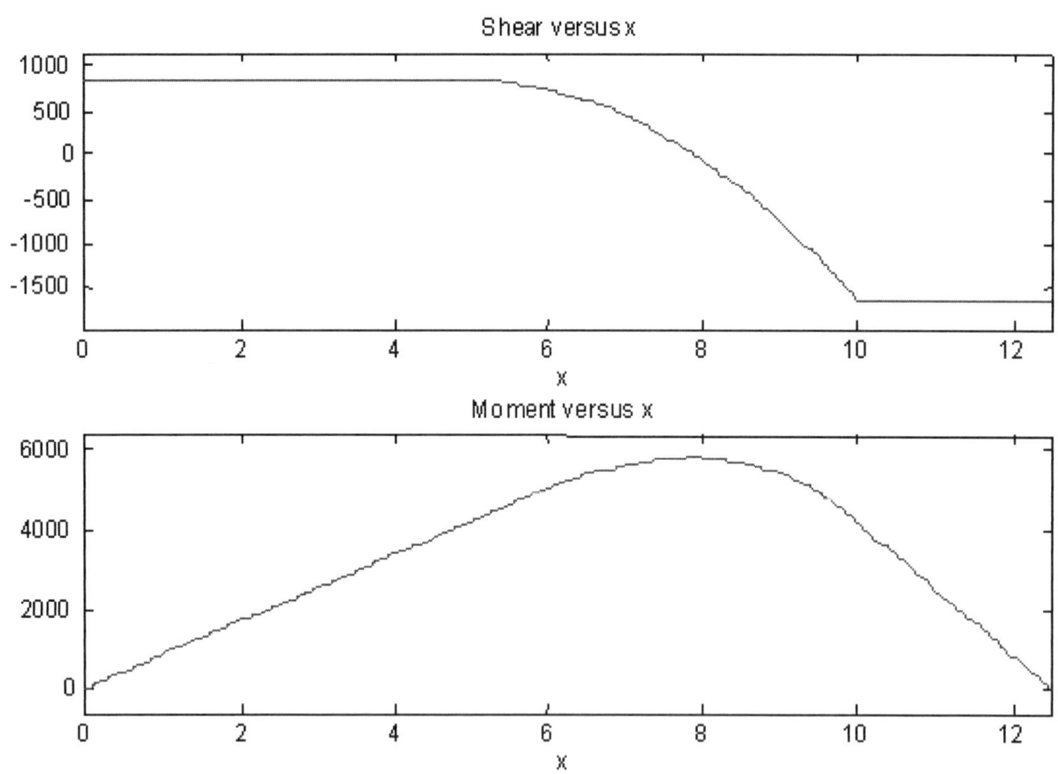

Cables

Problems involving cables loaded by concentrated forces result in systems of simultaneous nonlinear equations. MATLAB solves these equations by the use of the `fsolve` command, if available, and by the short version of it, called `mnewt`, described earlier in Chapter 2 and listed in the appendix.

Sample Problem 8.8

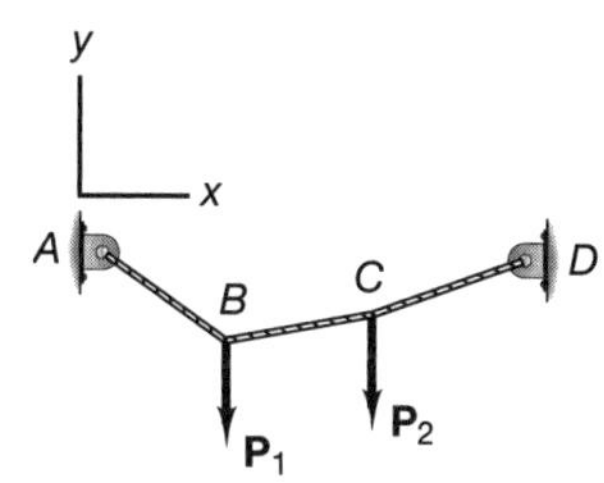

For the cable shown in the figure at the right, $P_1 = 750$ N, $P_2 = 400$ N; and the components of the distance between the supports are $AD_x = 30$ m and $AD_y = 0$ m. The design requires that the loads be equally spaced at $AB_x = 10$ m, $BC_x = 10$ m, and $CD_x = 10$ m. Determine the tensions in the sections of the cable if a cable of length 33 m is to be used.

First, create an m-file with the equations to be solved, saving it as `S8pt8_fcn.m`:

```
function q=S8pt8__fcn(p,L,P1,P2,ABx,BCx,CDx,ADy)
T1=p(1);T2=p(2);T3=p(3);
alpha1=p(4)*pi/180;
alpha2=p(5)*pi/180;
alpha3=p(6)*pi/180;

q=zeros(6,1);
q(1)=-T1*cos(alpha1) + T2*cos(alpha2);
q(2)=-T1*sin(alpha1) + T2*sin(alpha2) - P1;
q(3)=-T2*cos(alpha2) + T3*cos(alpha3);
q(4)=-T2*sin(alpha2) + T3*sin(alpha3) - P2;
q(5)=ABx*tan(alpha1) + BCx*tan(alpha2) + CDx*tan(alpha3) - ADy;
q(6)=sum([ABx BCx CDx]./cos([alpha1 alpha2 alpha3])) - L;
```

Next, construct an m-file that calls the previous commands and contains the parameters for $L = 33$ m.

SAMPLE PROBLEM 8.8: L = 33 M (CABLE LENGTH)

```
% Sample Problem 8.8 with L = 33. First enter the required constants
L = 33; P1 = 750; P2 = 400; ABx = 10; BCx = 10; CDx = 10; ADy = 0;
% put initial guesses into a column vector:
%      T1          T2          T3     alpha1  alpha2  alpha3     (DEGREES!)
IC = [1500;  1500;  1500;   -30;       10;          20 ];
% next call the Newton-Raphson routing, either fsolve or mnewt
soln = fsolve('S8pt8_fcn',IC,[],[],L,P1,P2,ABx,BCx,CDx,ADy);
% display the results
fprintf(['\nFor L = " num2str(L) ", the solution is\n'])
fprintf('T1       = %8.3f\n',soln(1))
fprintf('T2       = %8.3f\n',soln(2))
fprintf('T3       = %8.3f\n',soln(3))
fprintf('alpha1   = %8.3f\n',soln(4))
fprintf('alpha2   = %8.3f\n',soln(5))
fprintf('alpha3   = %8.3f\n',soln(6))
```

Saving this file (S8pt8_L33.m) and executing it yields the following solution:

```
For L = 33, the solution is
T1     = 1207.685
T2     = 1034.893
T3     = 1150.799
alpha1 =  -31.629
alpha2 =    6.473
alpha3 =   26.677
```

After the solution has been set up in an m-file, you can investigate the effect of parameters like the length of the cable. If the cable length was changed to 31 m, open the file S8pt8_L33.m and make the change. Then save your changes and run the file again by entering S8pt8_L31.m after the prompt to obtain the following solution:

SAMPLE PROBLEM 8.8: L = 31 M

```
For L = 31, the solution is
T1     = 1930.422
T2     = 1827.301
T3     = 1895.353
alpha1 =  -19.152
alpha2 =    3.661
alpha3 =   15.819
```

Sample Problem 8.9

A suspension bridge is designed to support a horizontal uniform load of 1000 N/m and span a distance of 50 meters. The distance from the lowest point on the main cable to the support on the left end is 10 meters, and that from the lowest point to the support on the right end is 12 meters. Determine the maximum tension in the cable.

A system of nonlinear algebraic equations can also be solved symbolically for certain types of nonlinearities. The symbolic solution seeks *all* the solutions and does not require an initial guess. The symbolic solution can also be converted to a numerical one if all parameter values are available. Establish the origin of the coordinate system at the lowest point on the cable, and determine x_L (the distance to the left support) and x_R (the distance to the right support) from the equation of a parabola, which is solved symbolically. Note that two solutions result, as should be expected in a nonlinear equation with quadratic terms. The second solution is the one of interest here. The code is as follows:

COMPUTATIONAL SOLUTION: SAMPLE PROBLEM 8.9

```
EDU>yL=10;yR=12;span=50;w=1000;
EDU>syms xL xR q To
EDU>eq1=xL+xR-span;
EDU>eq2=yL-q*(xL^2)/2;
EDU>eq3=yR-q*(xR^2)/2;
EDU>eq4=To-w/q;
EDU>S=solve(eq1,eq2,eq3,eq4)
S =
    To: [2x1 sym]
     q: [2x1 sym]
    xL: [2x1 sym]
    xR: [2x1 sym]
EDU>Ton=double(S.To),qn=double(S.q),xLn=double(S.xL),xRn=double(S.xR)
Ton=
1.0e+07 *
    1.3722
    0.0028
qn =
    0.0001
    0.0351
```

```
xLn =
  -523.8613
   23.8613
xRn =
   573.8613
    26.1387
```

Note that the first solution is not used, as it produces a negative value for one of the distances. So it is the second solution in each solution vector that is of interest here. Accordingly, we use q = 0.0351 and T_o = 28,470 to plot the tension in the cable:

```
EDU>x=-24:1:27;q=0.0351;
EDU>T=(2.847*10^4)*sqrt(1+(x*q).^2);
EDU>plot(x,T)
EDU>title('tension versus x')
```

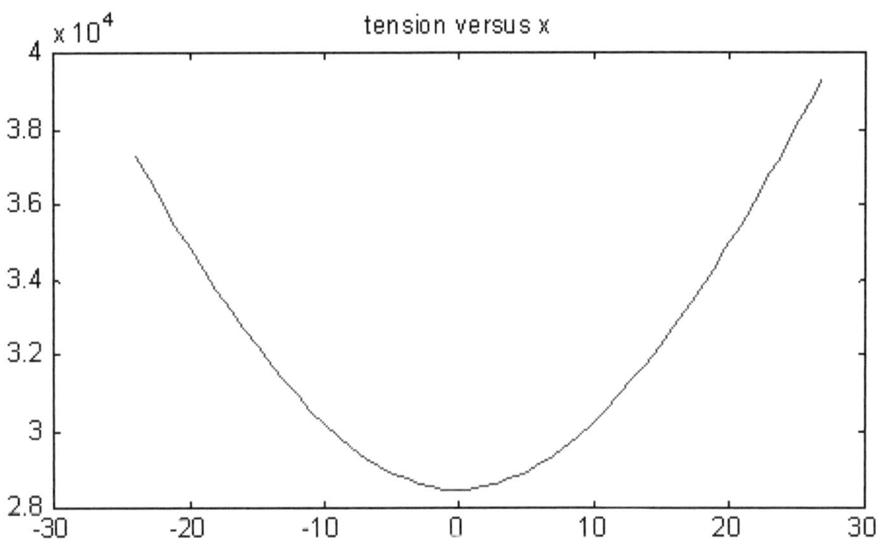

Sample Problem 8.10

Twenty meters of telephone cable with a weight of 40 N/m is supported by two poles of equal height placed 18 meters apart. Determine how low the cable drops and the maximum tension in the cable.

In this problem, a catenary cable of length 20 m spans a distance of 18 m and has a weight of 40 N/m. The value of q is found by solving the associated transcendental equation, as indicated in the code that follows. First, an m-file is created with the function defining q:

```
function y=sp8pt10(x)
y=sinh(9*x)-10*x;
```

Next, the command window is used to call this function into `fsolve` and plot the catenary:

```
EDU>q=fzero('sp8pt10',.1)                % use fzero to solve for q near 0.1
q =
   0.0893
EDU>x=-9:.1:9;y=(1/q)*(cosh(q*x)-1);     % define the range of x and the catenary
                                         % curve
EDU>plot(x,y),xlabel('x'),ylabel('y')    % plot the catenary
```

Hitting the return key results in the following plot:

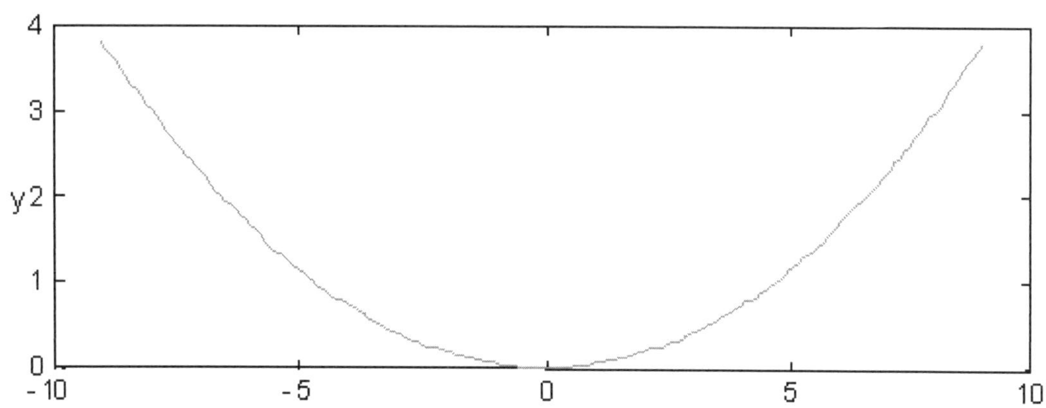

Next, consider defining and plotting the tension in the cable with the following commands:

```
EDU>To=40/q;
EDU>T=To*cosh(q*x);                                % define the tension as a function of x
EDU>plot(x,T),xlabel('x'),ylabel('T(x)')  % plot the tension as a function of z
```

Hitting the return key results in the following plot:

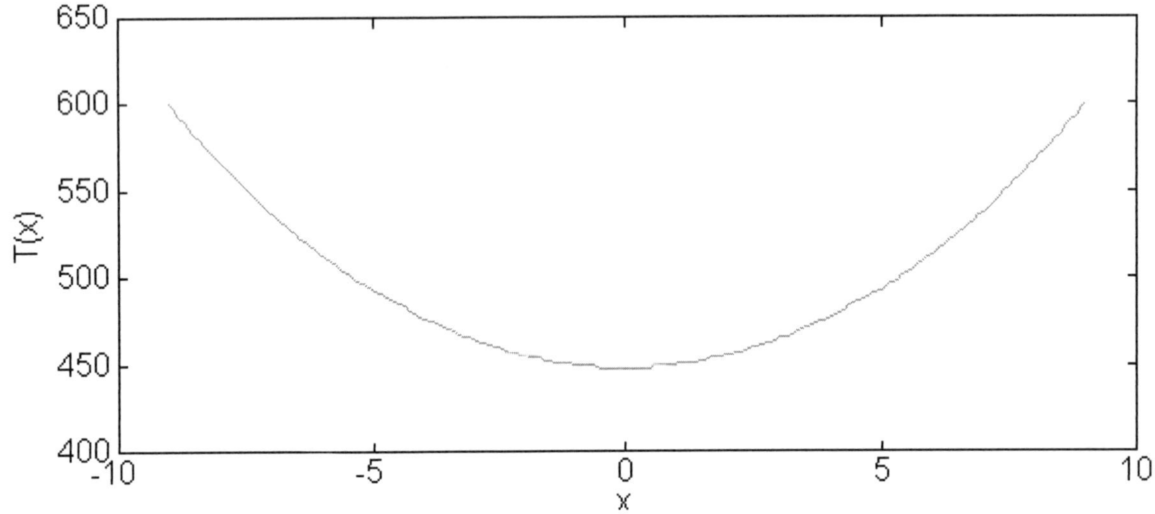

9

Friction

Coulomb friction, or dry friction, is an important concept. The coefficient of static friction is the ratio of the maximum friction force to the normal force between two contacting surfaces. Like any equilibrium problem, those having to do with friction involve a system of simultaneous equations, and MATLAB can be used to facilitate their solution.

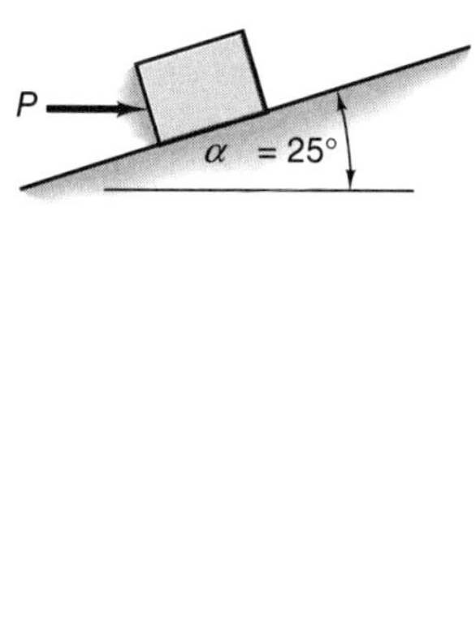

Sample Problem 9.2

Determine the range of values of the force P required to prevent the 400 N box shown at the right from slipping up or down the inclined plane if the coefficient of static friction between the box and the plane is 0.2.

The numerical work for Sample Problem 9.2 can easily be done using MATLAB. The two linear equations for P and N will be written in matrix notation and solved twice, once for a positive friction force and once for a negative friction force.

COMPUTATIONAL SOLUTION: SAMPLE PROBLEM 9.2

```
% this file solves sample problem 9.2
a=25*pi/180;                        % enter the angle and convert to radians
Cmin=[cos(a) .02; -sin(a) 1];       % enter the coefficient matrix with positive mu
Cmax=[cos(a) -.02; -sin(a) 1];      % enter the coefficient matrix with negative mu
Rs=400*[sin(a);cos(a)];             % enter the column matrix for the right side
min=Cmin\Rs; max=Cmax\Rs;           % compute the solutions in each equation
Pmin=min(1),Pmax=max(1)             % print out the min and max value of P
```

This file is saved under the name sp9pt2.m; enter the following to obtain the solution:

```
EDU> sp9pt2
Pmin =
  176.8735
Pmax =
  196.3543
```

Thus, the range of P for no slip is $97.4 < P < 293.9$ newtons. Note that this m-file computes the solution for both N and P in each case but by using the; command, these values are suppressed, as they are not of interest. The required information is in the first element of the solution vectors named min and max. The information is displayed by referring to the first element of each, i.e., min(1) and max(1).

Surface Plots

In Sample Problem 9.2, the force required to push the block up the incline may be computed as a function of the angle of incline (0 to 45°) and as a function of the coefficient of friction (0 to 0.7). The results can be displayed on an appropriate surface plot. The visualization provided by plotting the solution as a function of the angle and the coefficient of friction can help you to understand the relationship between these parameters. In this case, the relationship between P, μ, and θ may first be established symbolically by solving the equation $\mathbf{P} = c^{-1}\mathbf{R}$ in terms of symbolic versions of μ and θ. Solving the equation yields the functional form of the force P in terms of m and u, which can in turn be used to code a surface plot of P as a function of the two variables μ and θ.

To generate a surface plot in MATLAB, the commands mesh and meshgrid are needed. First, the range of values of interest for μ and θ is defined by using the colon notation. Next, these values are placed into a special grid format via the meshgrid command. The functional form of P is then defined in terms of the new meshgrid variables and plotted using either mesh or surf commands. We begin by computing the value of the vector **P** symbolically using the following m-file saved as friction.m:

```
syms t m
C=[cos(t) -m; -sin(t) 1];
R=400*[sin(t); cos(t)];
P=inv(C)*R;
P1=pretty(P(1))
```

The following command yields the form of the scalar P as a function of θ and μ:

```
EDU>friction
                          sin(t)                   m cos(t)
             -400 ------------------ - 400 ------------------
                    -cos(t) + m sin(t)        -cos(t) + m sin(t)
```

Next, we use this functional form of P to form a numerical expression to plot in mesh form:

FORCE VERSUS COEFFICIENT OF FRICTION AND ANGLE OF INCLINATION

```
d = 45*pi/180;               % define the constant d
t= linspace(0,d,15);         % define equally spaced values for the angle theta
m= linspace(0,0.7,15);       % define equally spaced values for the coefficient
                             % of friction
[T,M]= meshgrid(t,m);        % define the grid points followed by the function to
                             % be plotted
P1=-400*(sin(T)./(-cos(T) +M.*sin(T))-M.*cos(T)./(-cos(T) +M.*sin(T)));
mesh(T,M,P1)                 % this is the plot command for a mesh
xlabel('theta in radians')
ylabel('coefficient of friction')
zlabel('force')
```

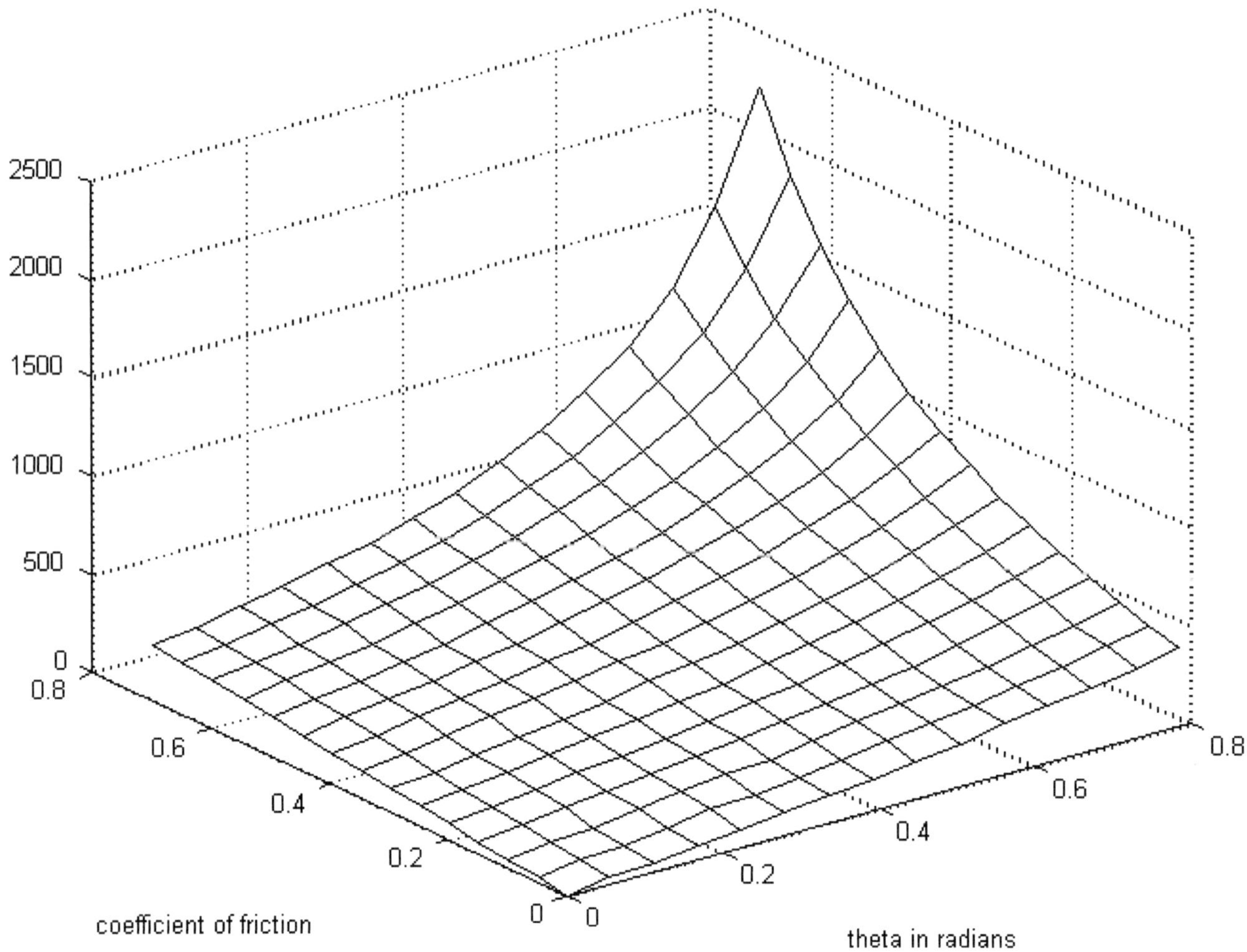

If the incline were at an angle of 45° and the coefficient of friction were 0.7, the horizontal force required to push the 400-newton box up the incline would be 2267 newtons. Analyses and surface plots of this type are used by engineers during the design stage.

Sample Problem 9.3

For the block shown at the right, determine the minimum force **P** as a function of θ for any angle of incline, α, to initiate motion.

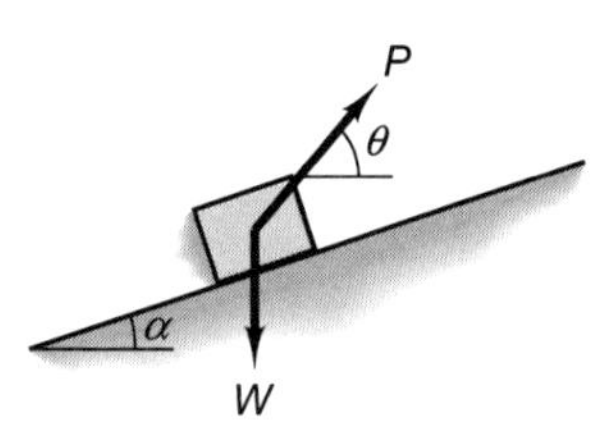

This sample problem is solved using MATLAB to determine the minimum force and the angle of application of the force for a specific angle that is necessary to initiate movement up the incline. First, create a symbolic version of the force expression in an m-file in order to differentiate the expression, and the plot both the force and its derivative:

```
syms t
W=200;a=25*pi/180;m=0.2;
P=W*(sin(a)+m*cos(a))/(cos(t-a)+m*sin(t-a));
DP=diff(P);
e=pi/2;
subplot(2,1,1)
ezplot(P,[0 e])
subplot(2,1,2)
ezplot(DP,[0 e]),grid
```

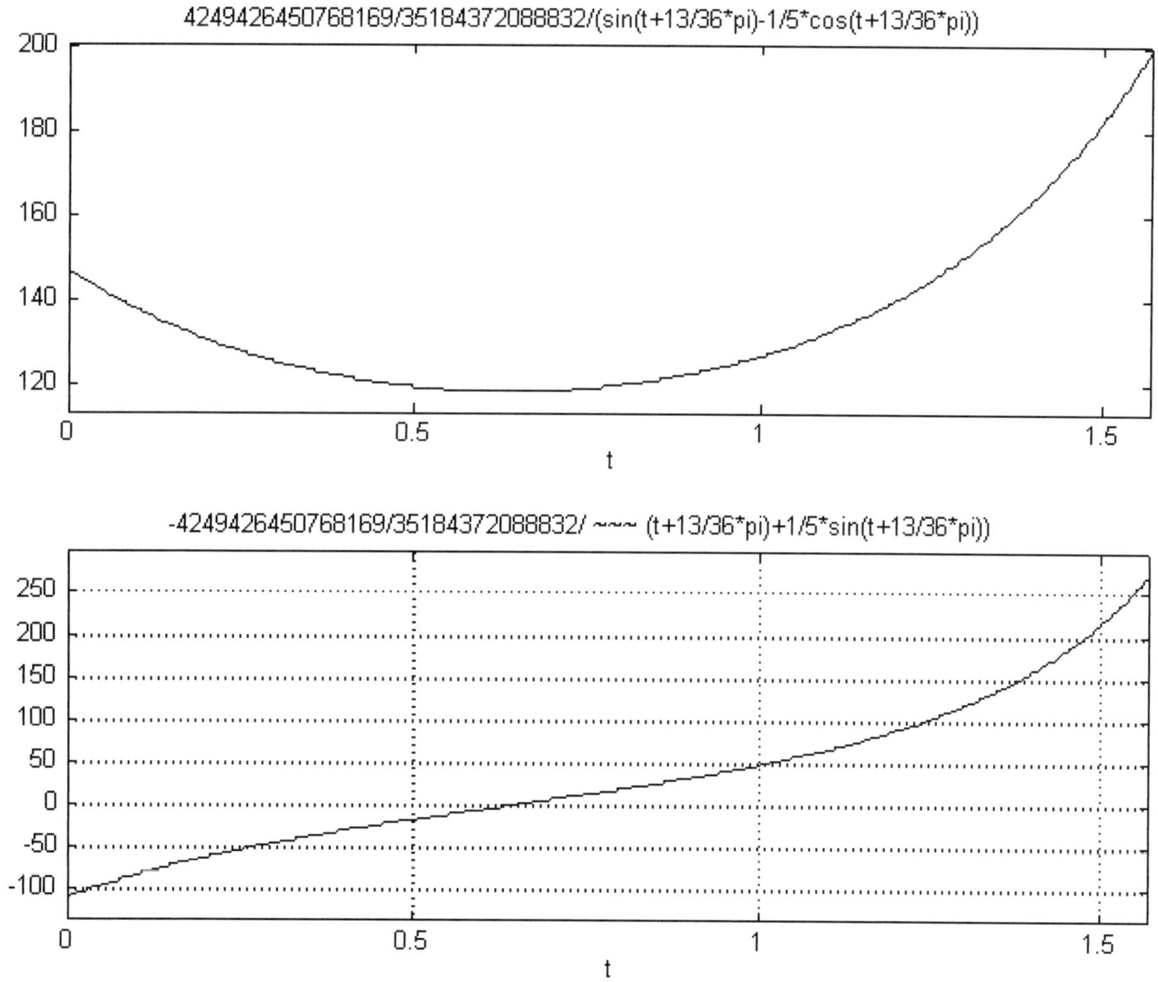

Examining the plot of the derivative, we see that it seems to cross zero around $t = 0.6$. The value of the symbolic derivative at $t = 0.6$ is obtained by entering the command subs, which evaluates a symbolic expression at a particular value. Note that increasing t a little eventually leads to the conclusion that $t = 0.6337$ is where the derivative vanishes:

```
EDU>double(subs(DP,t,.6))          % evaluate the derivative function at t =0.6
                                   % numerically

ans =
    -3.9982
EDU>double(subs(DP,t,.6337))
ans =
    -0.0033
EDU>double(subs(DP,t,.6336))
ans =
    -0.0151
```

To see what the value of the force is at $t = 0.6$ apply the subs command to the symbolic function P, as illustrated in the following command window:

```
EDU>double(subs(P,t,.6337))
ans =
   118.4306
EDU>double(subs(P,t,.6))
ans =
   118.4980
```

Thus, it appears that the smallest value of the force is about 118.5 N at $\theta = 36$ degrees.

Sample Problem 9.4

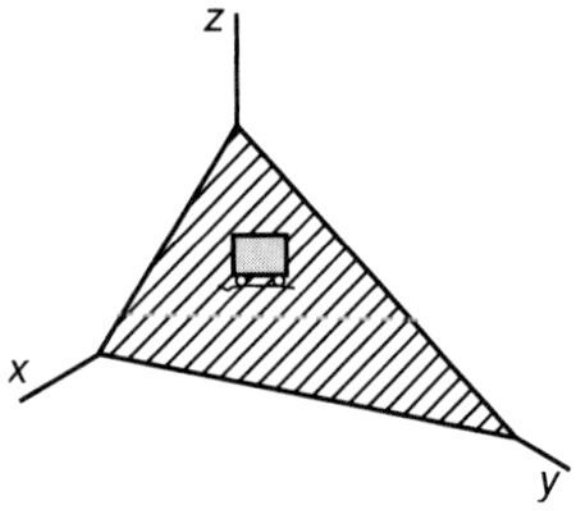

A block of mass m rests on an inclined plane, as shown in the diagram at the right. Determine the minimum coefficient of static friction required to keep the mass from sliding. The plane makes an angle α with the x-axis in the x–z plane and makes an angle β with the y-axis in the y–z plane.

This sample problem can be solved using MATLAB as a vector calculator for any numerical situation. Consider the case when $\alpha = 25°$, $\beta = 40°$, and the mass is 40 kg. Determine the minimum coefficient of friction to prevent slipping.

The following computes the coefficient of static friction:

NUMERICAL EXAMPLE: SAMPLE PROBLEM 9.4

```
% this file computes the coefficient of static friction for Sample Problem 9.4
a=25*pi/180; b= 40*pi/180; g=9.81;                % enter the angles and gravity
                                                  % values
m=[cos(a);0;-sin(a)]; q=[0;cos(b);-sin(b)];       % create the vectors n and q
n=cross(m,q)/norm(cross(m,q));                    % compute the vector n
W=[0;0; 40*g];                                    % define the vector W
Wn=dot(W,n)*n;        % calculate the component of W along the normal
N=-Wn; Wt=W-Wn;       % calculate N and the component of W tangent to the normal
f=-Wt;                % compute the friction force
mu=norm(f)/norm(N)    % compute the coefficient of static friction
```

This file is saved as sp9pt4.m, so that entering the following yields the desired coefficient of static friction:

```
EDU> sp9pt4
mu =
    0.9600
```

Wedges

The general solution of the four equations for the wedge system shown in Figure 9.10 can be solved symbolically using MATLAB. The coefficient matrix is entered, and the inverse of this matrix is identified by the command inv( ). The solution is then evaluated symbolically.

SYMBOLIC EVALUATION OF WEDGE SYSTEM: FIGURE 9.10

```
EDU>syms m1 m2 t m3 W
EDU>A=[0 -1 m2 0;0 -m1 1 0;1 0 -m2 -(sin(t)+m3*cos(t));0 0 -1 cos(t)-m3*sin(t)]
A =
[                0,               -1,               m2,                  0]
[                0,              -m1,                1,                  0]
[                1,                0,              -m2, -sin(t)-m3*cos(t)]
[                0,                0,               -1,  cos(t)-m3*sin(t)]
EDU>w=[0;W;0;0];
EDU>P=inv(A)*w;
EDU>pretty(P)
```

```
[   (-m2 cos(t) + m2 m3 sin(t) - sin(t) - m3 cos(t)) W  ]
[- ---------------------------------------------------- ]
[  cos(t) - m3 sin(t) - m1 m2 cos(t) + m1 m2 m3 sin(t) ]
[                                                       ]
[                         m2 W                          ]
[                      - ----------                     ]
[                        -1 + m1 m2                     ]
[                                                       ]
[                           W                           ]
[                      - ----------                     ]
[                        -1 + m1 m2                     ]
[                                                       ]
[                           W                           ]
[ ---------------------------------------------------- ]
[ cos(t) - m3 sin(t) - m1 m2 cos(t) + m1 m2 m3 sin(t)  ]
```

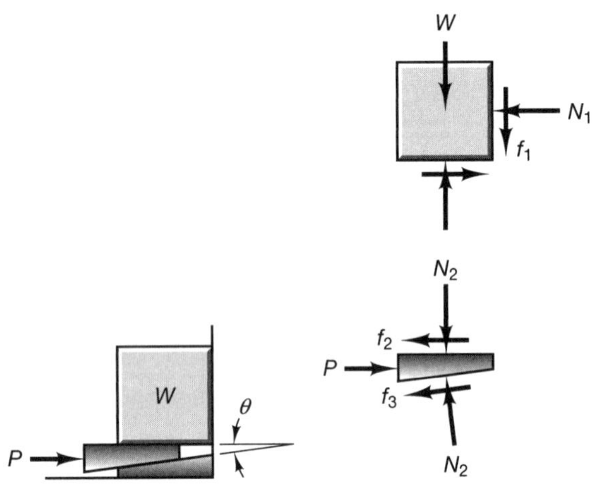

Figure 9.10

Dependence of P on the Wedge Angle *θ*. Consider the case when the coefficient of static friction is the same for all surfaces and is equal to 0.2. The applied force **P** is linearly related to the weight **W**. Therefore, the weight will be treated as a unit weight × 100 percent and the resulting normal forces and the applied force will be as a percent of **W**. The four linear equations will be solved using matrices. First, we establish an m-file that symbolically solves the equations of equilibrium in terms of the symbolic variable θ (denoted t in the file) and plots the first element. Note that the first component of the vector F in the code is the magnitude of the applied force, P, the second element is N_1, the third is N_2, and the last N_3:

```
syms t
m=0.2;
A=[0 -1 m 0; 0 -m 1 0; 1 0 -m -(m*cos(t)+sin(t));0 0 -1 cos(t)-m*sin(t)];
L=[0;100;0;0];
F=inv(A)*L;
P=F(1);e=20*pi/180;
ezplot(P,[0 e])
title('force versus wedge angle in radians')
```

Saving and executing this file yields the following plot:

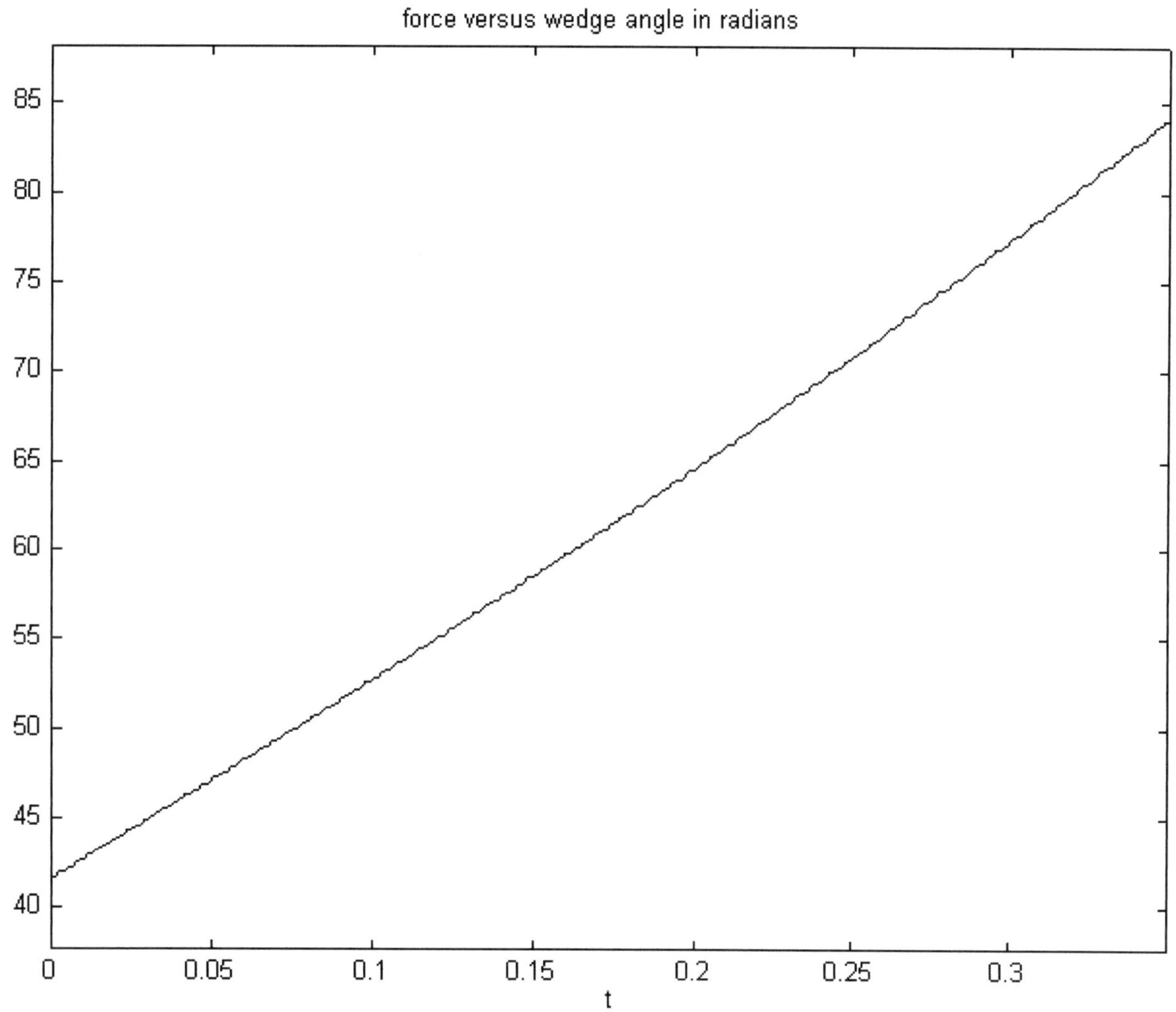

Next, several values of P are calculated at different angles using the `subs` command:

```
EDU>double(subs(P,t,1*pi/180))
ans =
   43.5643
EDU>double(subs(P,t,20*pi/180))
ans =
   84.1924
```

In this case, the wedge angle was varied from 1° to 20°, and the required wedge force went from 43.6% of the weight to 84.2% of the weight. The greater wedge angle would lift the weight higher than the smaller wedge angle would.

Dependence of P on μ in Percent of W. The dependence of the force **P** on the coefficient of friction is shown in the next m-file. A specific wedge angle of 10° is chosen for this example. Then the previous m-file is copied and the symbolic variable changed from θ to μ (denoted `m` in the file). The modified file is as follows:

```
syms m
t=10*pi/180;
A=[0 -1 m 0; 0 -m 1 0; 1 0 -m -(m*cos(t)+sin(t));0 0 -1 cos(t)-m*sin(t)];
L=[0;100;0;0];
F=inv(A)*L;
P=F(1);e=0.6;
ezplot(P,[0 e])
title('force versus the coefficient of friction')
```

Saving and executing the file yields the following plot:

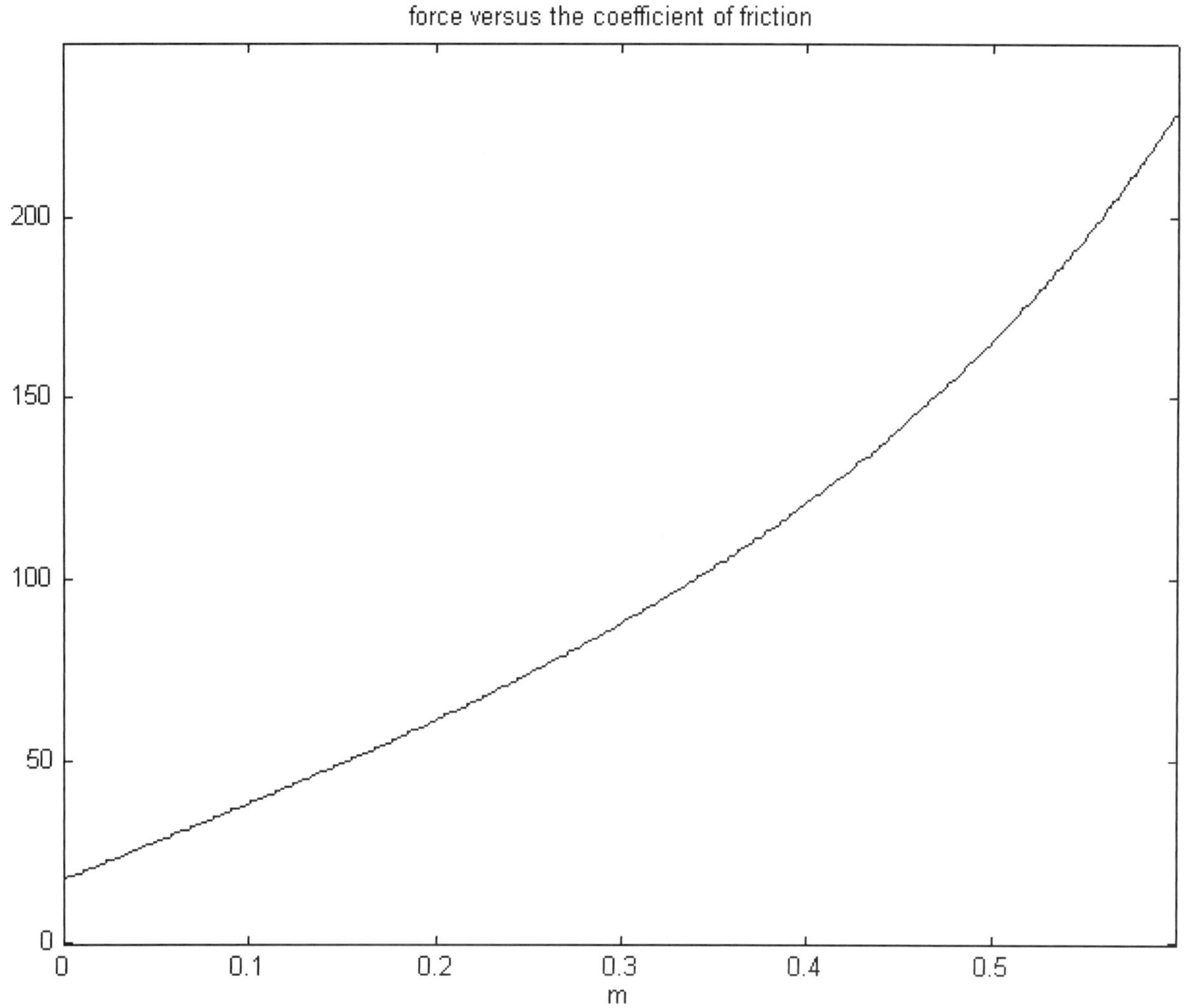

It is easily seen that wedges are very sensitive to the coefficient of friction between the contacting surfaces. For smooth surfaces, the force required to raise the weight is 17.6 percent of the weight, while with a coefficient of friction of 0.6, the force *P* is 229.4 percent of the weight. Although friction appears to be detrimental to the use of the wedge, some friction is necessary to hold the wedge in place.

Belt Friction

The next m-file plots the ratio of belt tensions for a belt-and-pulley system as the coefficient of friction changes for a fixed contact angle. Then the belt tension ratio is plotted as a function of the contact angle for a fixed coefficient of friction. This is followed by a surface plot illustrating the effect of varying both the angle and the friction coefficient at the same time.

RATIO BETWEEN TENSIONS VERSUS COEFFICIENT OF FRICTION

```
% this m file computes the tension ratios for various coefficients of friction
  in belts
b=pi/2; m=0:.05:1;   % set the belt angle beta and define the range of friction
                     % coefficients
R=exp(b*m);                    % caclulate e to the beta mu
plot(m,R)
xlabel('coefficient of friction')
ylabel('ratio')
title('ratio of T2 to T1')
```

Saving this m-file as `tbelt.m` and entering `tbelt` after the prompt in the command window results in the following plot:

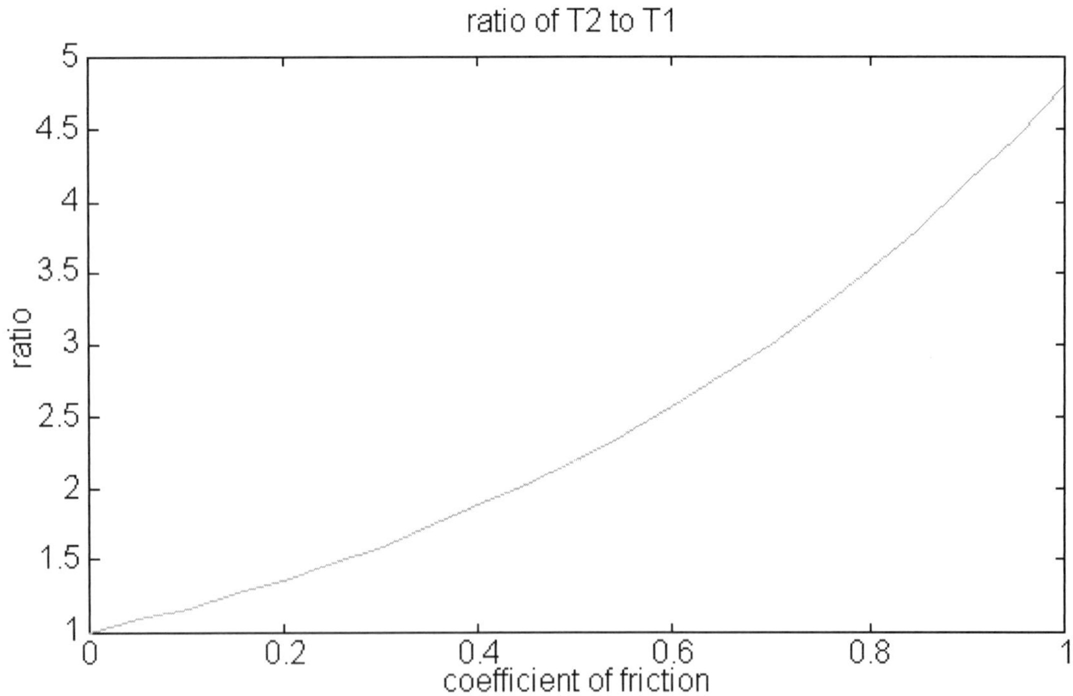

It is also of interest to investigate the effect of increasing the contact angle. A plot of the ratio of the tensions versus the contact angles varying from 0° to the angle obtained when the belt is wrapped around the drum two and one half times is generated by the following MATLAB code:

RATIO OF TENSIONS VERSUS THE ANGLE OF CONTACT

```
% this m file computes the tension ratios versus the angle of contact
m=0.4; b=0:.02*pi:5*pi;      % set the friction coefficient and define the range
                             % of belt angle beta
R=exp(b*m);                            % caclulate e to the beta mu
plot(b,R)
xlabel('angle beta')
ylabel('ratio')
title('ratio of T2 to Tl')
```

Saving this m-file as `cont.m` and entering `tcont` after the prompt in the command window results in the following plot:

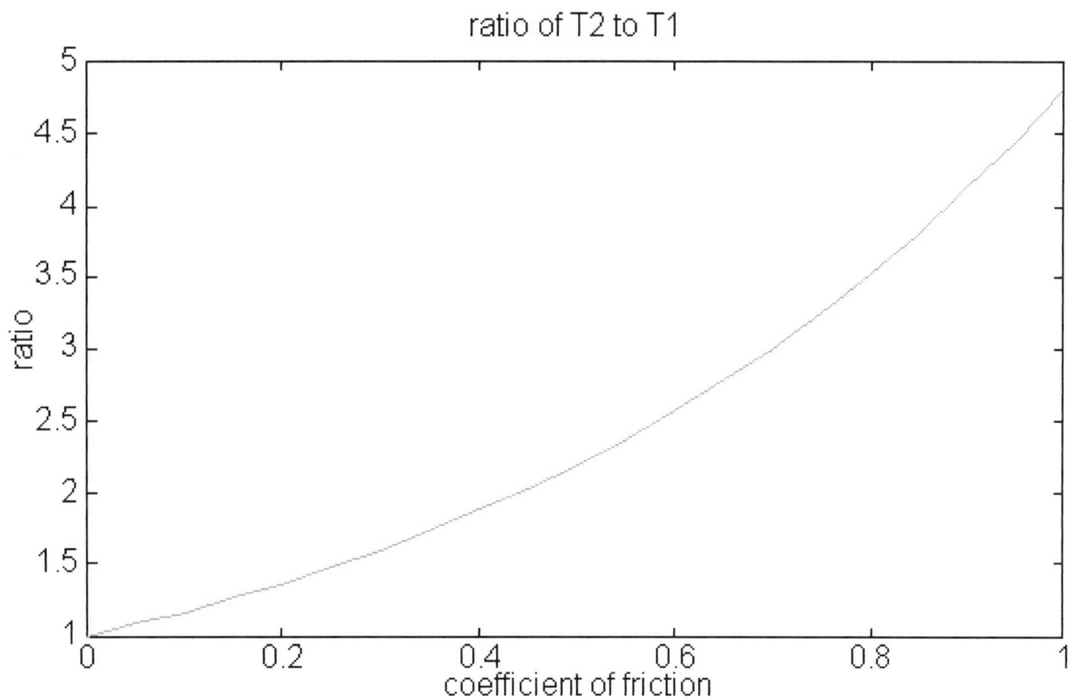

Next, a surface plot is made to show the full dependence of the belt friction on both the coefficient of friction and the contact angle β. This requires using the surface plot `mesh` and `meshgrid` as outlined earlier.

Ratio of Tensions Versus Coefficient of Friction and Contact Angle

MATLAB allows you to conveniently construct surface plots employing essentially the same sort of plot commands as are used in two dimensions. Here, however, the domain of the two independent variables must be the same and must be evaluated at equal increments. Hence, the `linspace` command is used to specify the range of values for each variable. The commands `mesh` and `meshgrid` prepare the data for plotting in a three-dimensional wire-mesh surface plot. This is illustrated in the following m-file:

```
% this plots the ratio of tensions versus angle and coefficient of friction
a=45*pi/180;
t=linspace(0,a,10); m=linspace(0,0.7,10);   % this sets evenly spaced data along
                                            % the x and y axis
[T,M]=meshgrid(t,m);                    % this prepares the x-y points (t-m) into a
                                        % matrix for plotting
Ratio=exp(T.*M);                        % defines the function to be plotted, note
                                        % the semi colon
mesh(T,M,Ratio)                         % defines the mesh and plots it (no semi
                                        % colon)
xlabel('theta in radians')              % the axis along x, which is t in this case
ylabel('coefficient of friction')       % the axis along y, which is m in this case
zlabel('force in Newtons')              % the axis along z, which is the function
                                        % being plotted
```

This code is saved in the m-file named `tf3d.m`, so that entering `tfz3d` after the prompt produces the following plot:

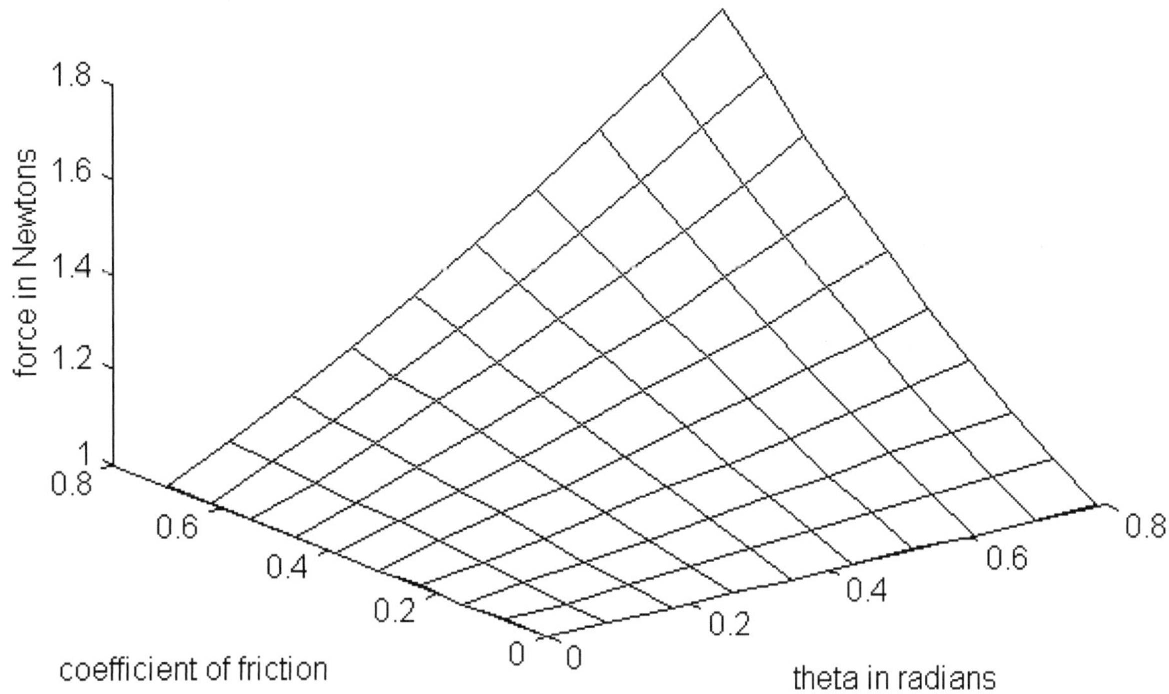

Surface plots are useful to see how an analysis depends on the variables involved. The plot shown was done for coefficients of friction from 0 to 0.7 and for values of theta from 0 to 45 degrees.

10

Moments of Inertia

Principle Second Moments of Area

Many of the integrals arising in the computation of the second moment of an area can be evaluated symbolically on MATLAB. Consider, for example, the second moment of the circular area shown in Figure 10.5 about its centroidal axis. MATLAB code for calculating this moment is shown in the Computational Window 10.1.

COMPUTATIONAL WINDOW 10.1

```
EDU>syms r R t
EDU>f1=(r^3)*sin(t)^2;
EDU>f2=int(f1,r,0,R); A=int(f2,t,0,2*pi);
EDU>f2=int(f1,r,0, R); A=int(f2,t,0,2*pi);
EDU>pretty(A)
                                                 4
                                        1/4 pi R
```

It is useful to plot the relationships between of the second moment of area and the product moment of area on the coordinate system rotation angle. Computational Window 10.2 produces a plot of equations (10.33) and (10.34).

COMPUTATIONAL WINDOW 10.2: SECOND MOMENT OF AREA

```
Ixx=10;Iyy=5;Ixy=0;                               % define the constants
b=0:0.1:2*pi;                                     % define the range of values of beta
pIxx=(Ixx+Iyy)/2+(Ixx-Iyy)*cos(2*b)/2+Ixy*sin(2*b);    % define the prime
                                                        % inertia Ix'x'
pIxy=-(Ixx-Iyy)*sin(2*b)/2+Ixy*cos(2*b);  % define the prime inertia Ix'y'
plot(b,pIxx,'+',b,pIxy,'*')                       % plot both functions
xlabel('beta')
title('inertias versus the angle beta')
```

This file is saved under the name 10pt2.m, so that entering 10pt2 after the prompt returns the following plot:

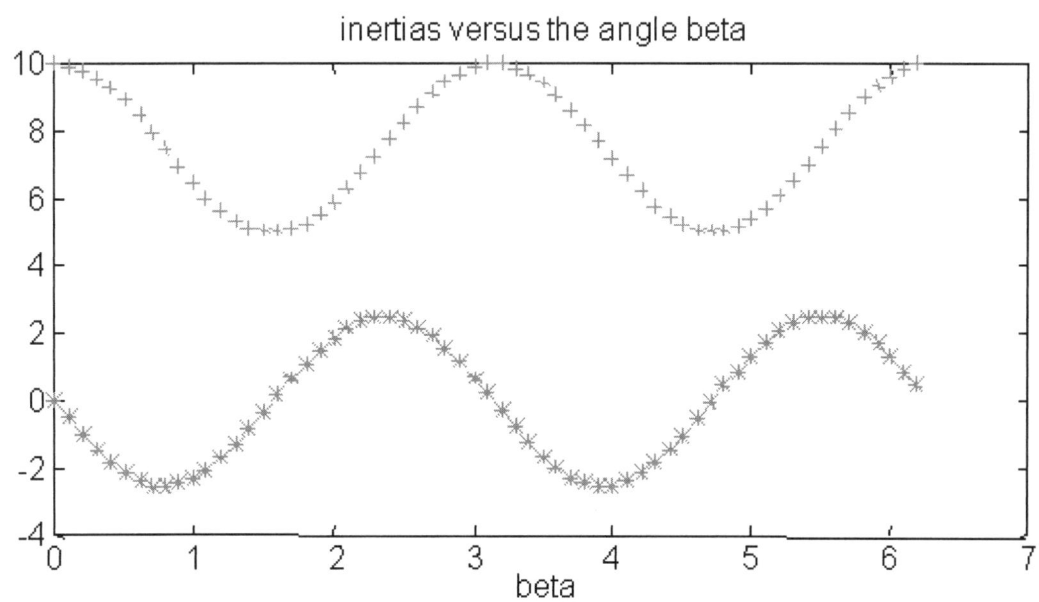

Sample Problem 10.5

For the area shown to the right, find the centroid, and the principal centroidal axes, and the value of the second moment of area about these axes.

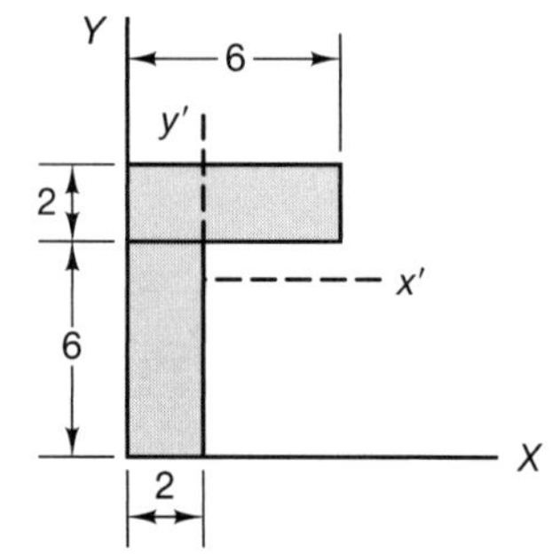

This preceding m-file, used to plot the inertia, can do the same for Sample Problem 10.5. Just open the file and change the values of I_{xx}, I_{xy}, and I_{yy} to obtain the new values. The following plot results from typing `10pt2` after the prompt:

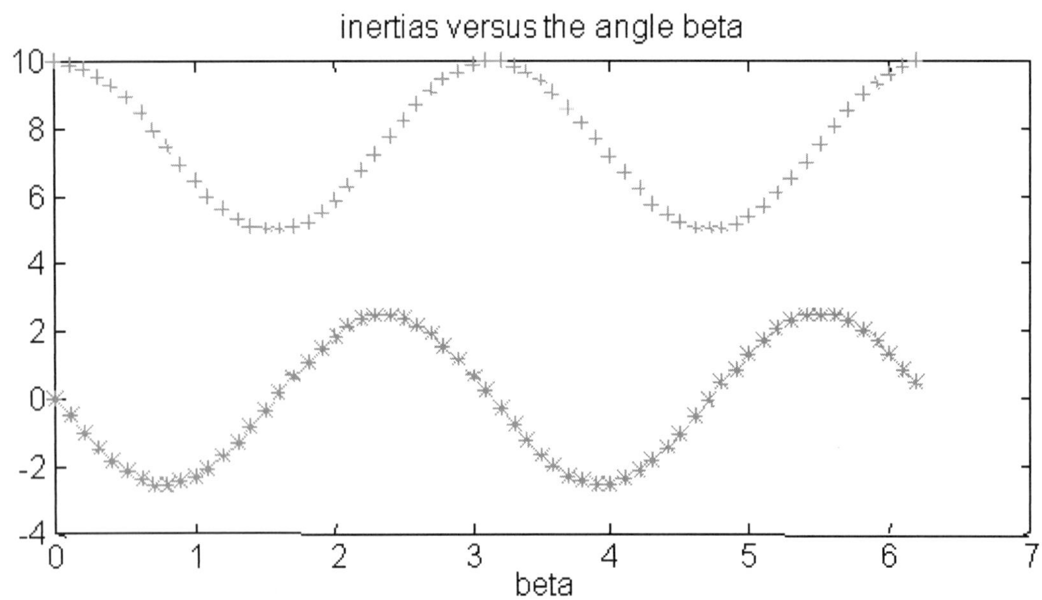

Eigenvalue Problems

An eigenvalue problem is presented in Section 10.10 of *Statics*. MATLAB can find the eigenvalues and eigenvectors of any square matrix using the simple commands `eig(M)` or `[U,D] =eig(M)`, where M is a square matrix with numerical entries. For small-order matrices, the eigenvalues and eigenvectors can also be found symbolically. The command `eig(M)` returns a column vector with the eigenvalues of the matrix **M** listed as its elements. The command `[U,D] =eig(M)` returns the eigenvectors as the columns of the matrix **U** and a diagonal matrix **D** containing the eigenvalues.

Sample Problem 10.5 is re-solved with the following m-file using the command `[U,D] =eig(M)`:

```
% Sample Problem 10.5 using matrix methods
I=[136 -48; -48 64];                 % enter the second moment of area matrix
[U,D]=eig(I)                         % compute its eigenvectors and eigenvalues
b1=atan(U(1,1)/U(2,1))*180/pi        % compute the angle form the x axis of the 1st
                                     % eigenvector
b1=atan(U(1,2)/U(2,2))*180/pi        % compute the angle form the x axis of the 2nd
                                     % eigenvector
```

This file is saved as `sp10pt5.m`, so that entering `sp10pt5` after the prompt results in the following solution:

```
EDU>sp10pt5
U =
   -0.8944   -0.4472
    0.4472   -0.8944
D =
   160     0
     0    40
b1 =
  -63.4349
b1 =
   26.5651
```

Here, b1 and b2 are the angles, in degrees, made with the *x*-axis for each of the two vectors, and the principal values are 160 and 40.

Note that `U(1,1)` is the first element of the first eigenvector and that `U(2,1)` is the second element of the first eigenvector. Looking at the matrix **U**, we see that the first column is the eigenvector corresponding to $\lambda_1 = 160$ and the second column is the eigenvector corresponding to $\lambda_2 = 40$. This solution is 180 degrees from the answer obtained in Sample Problem 10.5, but the axis is still the same.

The eigenvectors are perpendicular, which you can check by entering the following code:

```
EDU>dot(U(1,:),U(2,:))      % take the dot product of the first column with the
                            % second
ans =
     0
```

11

Virtual Work

The work done by a particle is equal to the integral of the dot product of the force acting on the particle and the differential change in the position vector. MATLAB can be used symbolically or numerically to evaluate some of these integrals.

Sample Problem 11.3

Consider a path defined by $\hat{\mathbf{t}}[\theta(s)]$, where $\theta(s) = 1 + s^2$. It can be shown that this path is a spiral in the plane. Plot the path of the particle in the plane and determine the work done by a constant force in the x-direction of 50 N as the particle moves 4 meters along the path.

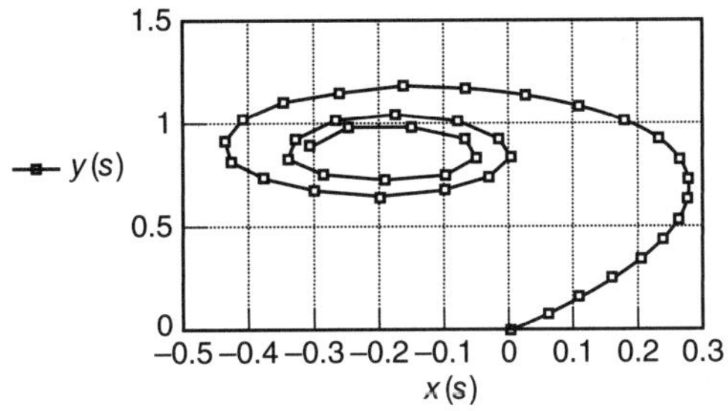

curve defined by $\theta(s) = 1 + s^2$

The integrals in Sample Problem 11.2 must be evaluated numerically, as no exact forms exist. An examination of the special functions found in any standard table of integrals reveals that the integrals in this sample problem will involve Fresnel sine and cosine integrals. First, two function files with the integrands must be prepared. The first m-file contains the integrand for y, saved as `s11.2.m`:

```
function y=s112(u)
y=sin(1+u.^2);
```

The next m-file contains the integrand for x, saved as `c112.m`:

```
function y=c112(u)
y=cos(1+u.^2);
```

The following m-file computes the integrals and sets up a plot of the functions:

SAMPLE PROBLEM 11.3

```
s=0:0.1:4;
n=length(s);
for i=1:n;
      xs(i)=quad8('c112',0,s(i));
      ys(i)=quad8('s112',0,s(i));
      end;
plot(s,xs,'*',s,ys,'+')
title('x(s) denoted by * and y(s) denoted by + versus s
```

This file is saved as `sp11pt2.m`, so that entering `sp11pt2` after the prompt produces the following plot:

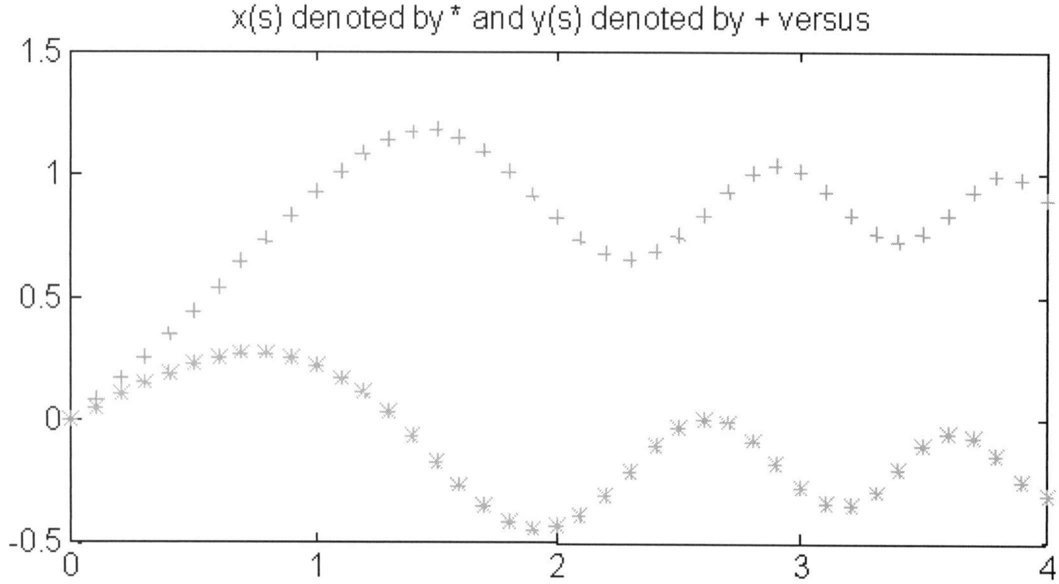

A plot of $x(s)$ versus $y(s)$ can be obtained by entering

```
EDU> plot(xs,ys), title('x(s) versus y(s)')
```

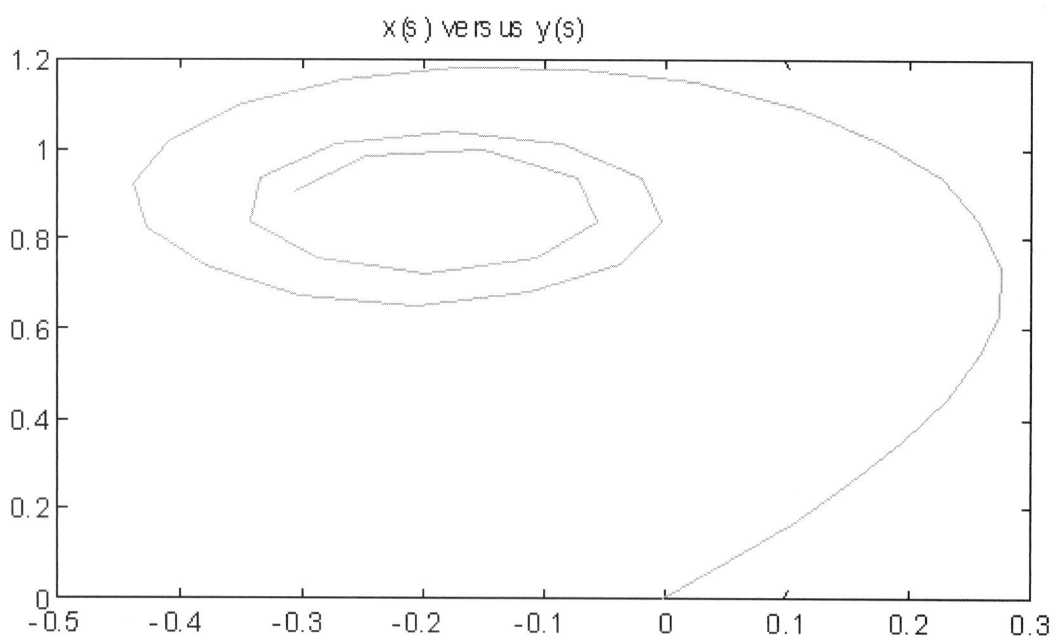

Next we compute the work done by a constant force of 50 N acting in the x direction by first setting up the integrand in the following m-file, saved as `s50n.m`:

```
function y =s50n(u)
y=50.*cos(1+u.^2);
```

Then we create the following m-file to perform and plot the integration:

```
s=0:0.1:4;
n=length(s);
for i=1:n;
        xs(i)=quad8("s50n',0,s(i));
        end;
plot(s,xs)
title('U(s) versus s')
xlabel('s')
```

Entering the name of the file after the prompt results in the following plot:

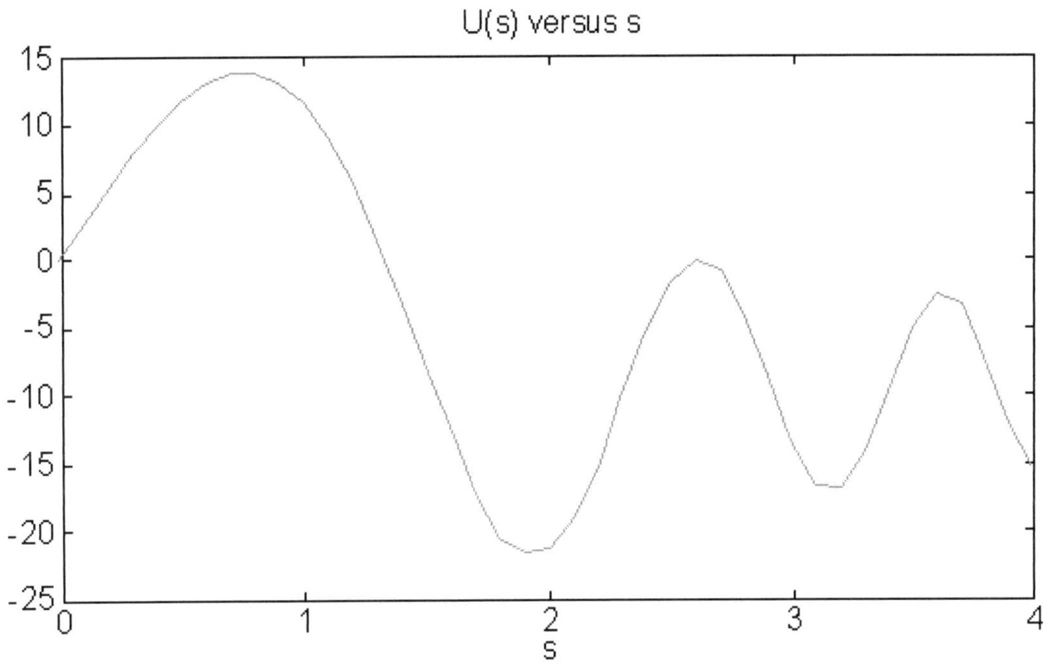

Notice that the force does both positive and negative work as it moves along the path, and at 4 meters the total work done is negative.

Sample Problem 11.4

Consider a child of 20-kg mass sliding down a slide in a playground with a coefficient of friction of 0.3. The equation for the slide path can be considered to be parabolic in nature; the initial angle of the slide is $-60°$, and the final angle is 0°. If the length along the slide is taken to be 1 unit, the total work can be determined. *Note that, since the child, modeled as a particle, is moving along a curved path in space, the normal force would also contain an inertial term equal to the product of the mass and the velocity squared, divided by the radius of curvature of the path in space. The problem becomes nonlinear if this inertial term is included; the problem then becomes one of dynamics, not statics. Here, we consider the inertial term negligible, and the normal force will balance the gravitational attraction.*

Integrals similar to those of Sample Problem 11.2 arise in Sample Problem 11.4, and their numerical solution is shown. In this problem, a slide has an initial angle of 60 degrees with the horizontal and makes a final angle of 0 degrees. Note that the path of the slide is parabolic.

First, we establish the following two m-files, which define the relevant integrands for x and y, to be used in the file that calls the integration commands:

```
function y= c114(u)
y=cos(-(1-u.^2).*pi/3);
```

```
function y=s114(u)
y=sin(-(1-u.^2).*pi/3);
```

Next, an m-file is created to integrate repeatedly and computes each integral as a function of the limit of integration using the `quad8` command and a for loop. Note that this loop calls the two functions `s114` and `c114`. The code of the m-file is as follows:

```
s=0:0.1:1;
n=length(s);
for i=1:n;
       xs(i)=quad8('c114',0,s(i));
       ys(i)=quad8('s114',0,s(i));
       end;
plot(xs,ys)
title('x(s) versus y(s)')
```

Saving this file (as `sp11pt4`) and executing it yields the following plot:

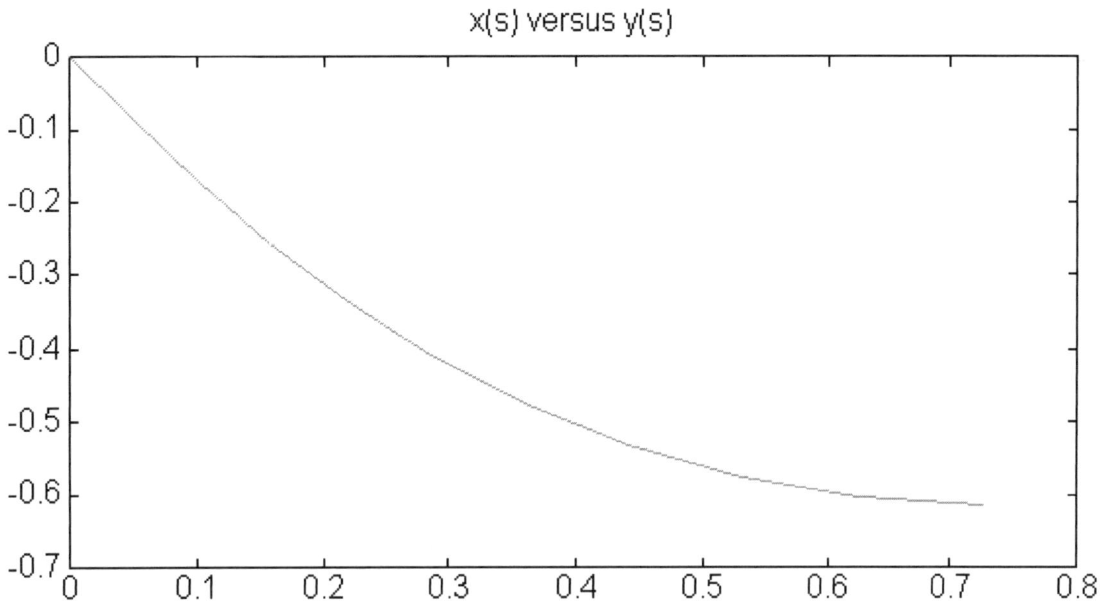

The total work done by a slide with a normalized length of unity can also be calculated, as illustrated in the next two files. The first m-file establishes the integrand and the second performs the integration:

```
function y=U114(u)  % this defines the integrand for the work calculation
m=20; mu=0.3; W=m*9.81; % enter the coefficient of friction & mass
y=-W*sin(-(1-u.^2).*pi/3)-mu*W*cos(-(1-u.^2).*pi/3);
```

```
% this computes the work done for a slide
s=0:0.1:1;     % define the points at which the integral will be evaluated
n=length(s);   % determine the number of points of evaluation
for i=1:n;
      U(i)=quad8('U114',0,s(i));
end;
plot(s,U), xlabel('s'),title('Work U(s) versus s')
```

Entering the name of this last file (u114) after the prompt results in the following plot:

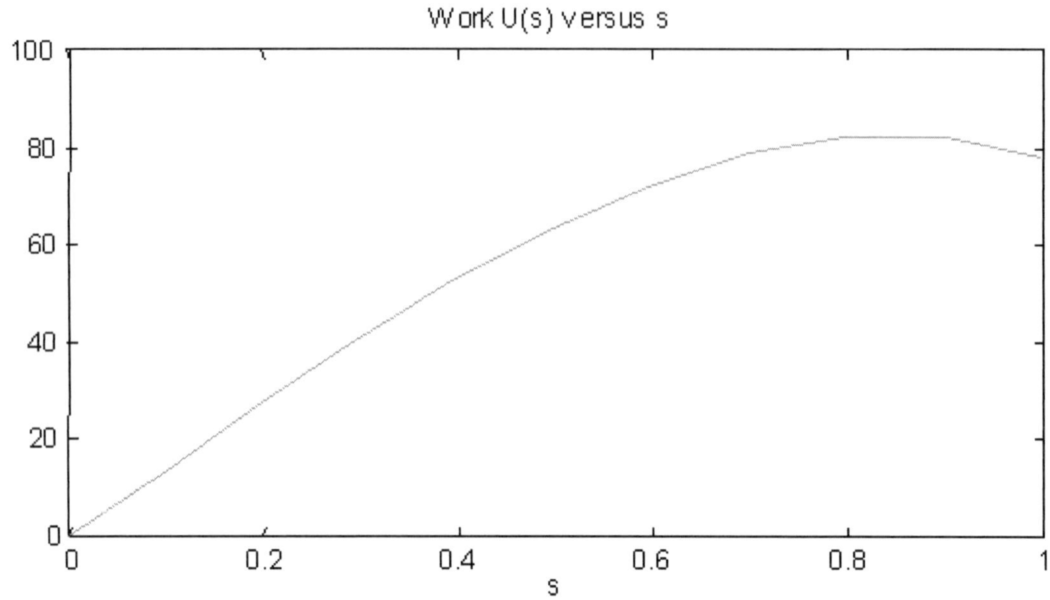

The final value is obtained by simply calling for it in the command window as follows:

```
EDU>U(n)
ans =
   77.7408
```

To determine the total work done, multiply $U(1)$, stored in U(n) here, by the total length of the slide. This length may be determined by the height of the slide, which is about 0.6s. Note that this information can be used in a dynamic analysis to predict the velocity of the child at the bottom of the slide.

Another variation of the preceding problem is to examine the effect of the coefficient of friction. As the coefficient of friction increases, more negative work will be done and the child will slow down. When the total work done is zero, the child will stop, whether or not he or she has reached the bottom. To see this, the total work is recomputed with a new coefficient of friction $\mu = 0.9$. To recompute the total work done in this case, just change the value of mu in the m-file U114.m, and run the m-file u114.m again to get the following plot:

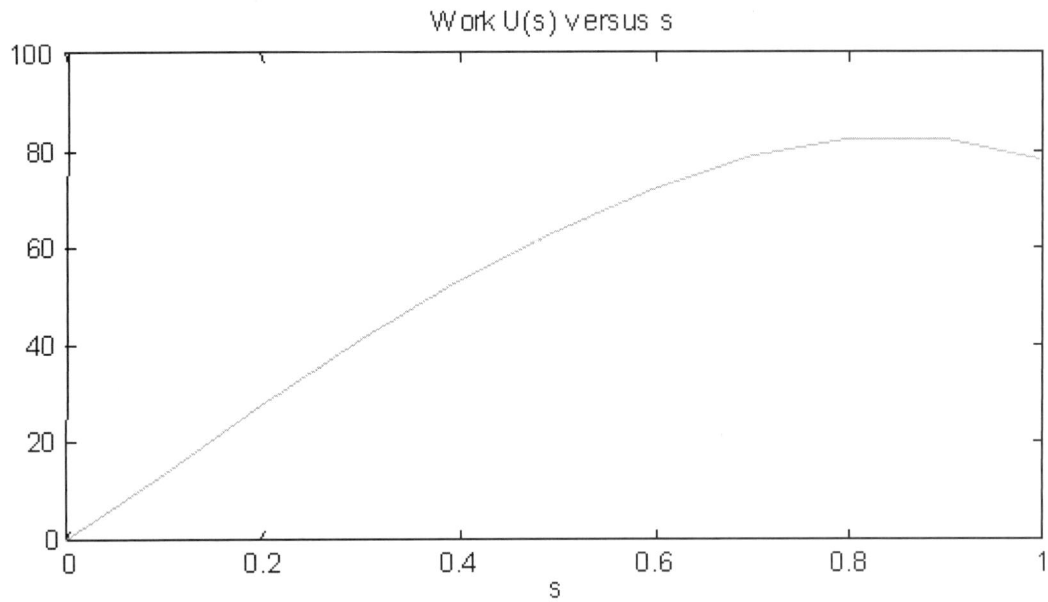

From this plot, it can be seen that the work becomes zero around $s = 0.95$ and is negative thereafter. Thus, the child sliding down the slide will stop after traveling only 95 percent of the way along the slide. This problem is used again in the *Dynamics* text, where it is shown that the velocity of the child is related to the work done. In this case, the child's maximum velocity will be about $s = 0.5$.

Sample Problem 11.7

Find the critical points of $f(x,y) = x^2 + 3xy + y^2$, and determine whether there is an extremum at these critical points.

A surface plot of the function in Sample Problem 11.7 shows the general shape of the function. Surface plots may be generated in MATLAB by using the `mesh` and `meshgrid` commands, as illustrated at the end of Chapter 9, and used in the next m-file. If you do not like the orientation of your plot, you can use the `view` command to reorient the plot. Type `help view` for directions on reorienting the plot. The following code will produce the desired plot!

SURFACE PLOT: SAMPLE PROBLEM 11.7

```
% this file makes a surface plot of the function of Sample Problem 11.7
x=linspace(-2,4,20);                % defines the x-axis values
y=linspace(-2,4,20);                % defines the y-axis values
[X,Y]=meshgrid(x,y);                % sets up the x-y grid for plotting
F=X.^2+3*X.*Y+Y.^2;        % defines the function to be plotted (note of element
                           % operations)
mesh(F)                             % plot F vs x and y
xlabel('x'),ylabel('y'),zlabel('F(x,y)')
```

This code is saved in the file `sp11pt7.m`, so that entering `sp11pt7` after the prompt produces the following plot:

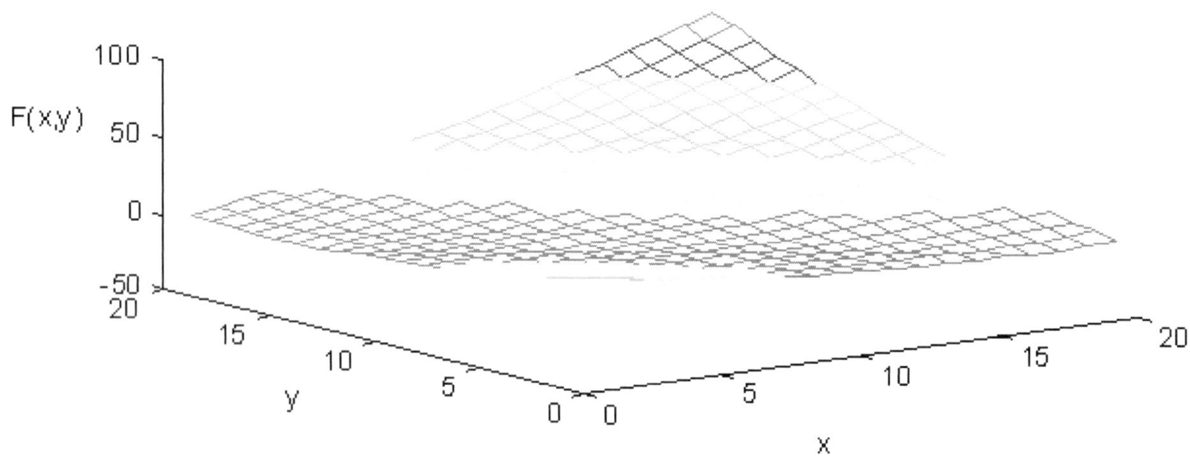

Summary

This supplement has shown how to use MATLAB to solve problems presented in the text *Engineering Mechanics: Statics*. Many of these problems may be solved by hand, and MATLAB reduces the labor of numerical calculations. The nonlinear problems cannot be solved without the use of some computational aid and are examples of the increased capabilities of analysis when computational software is used. This supplement is not meant to replace the manual for MATLAB or to show all the features of the software. Note that MATLAB is a tool for computation and is not a statics software package. That is, all problems must be modeled and the equations of equilibrium written before any software is used.

Acknowledgments

The authors extend their sincere thanks to Scott Gray and Cole Brooking, of The MathWorks, Inc. Cole was kind enough to provide the program in the appendix, as well as point out how to use the Heaviside function by referring to the `stepfun` command.

Appendix

Solution of Systems of Nonlinear Equations in MATLAB

The window that follows shows the code to enter for solving a system of nonlinear equations using a Newton-Raphson method for those who do not have the function `fsolve` available to them. The command `fsolve` is found in the Optimization Toolbox as an optional attachment to the professional version of the software. The commands written in the following window are a short version of the code for `fsolve` which should be suitable for the problems encountered in statics. The commands are designed to be used in much the same way that `fsolve` is used.

```
function a = mnewt(FUN,ntrial,guess,tolx,tolf,varargin)
%MNEWT Solve multiple nonlinear equations using Newton-Raphson method
%
%   MNEWT(FUN, ntrial guess, tolx, tolf, parameters)
%   FUN - name of the function to be tested, function returns
%      results in a column vector
%   ntrial - maximum number of trials
%   guess - initial guess
%   tolx - tolerance in the iterative step size
%   tolf - tolerance in function results
%   parameters - parameters passed directly to FUN
%      after the guess
% by Cole Brooking, The MathWorks, Inc. 7/3/97
for k = 1:ntrial
   beta = feval(FUN,guess,varargin{:});
   for i = 1:length(guess) % Compute partials
      guess2 = guess; delta = abs(guess(i)*1e-5);
      guess2(i) = guess2(i)-delta;
      alpha(:,i) = (feval(FUN,guess2,varargin{:}) - beta)./delta;
   end
   errf = sum(abs(beta)); if errf < tolf; break; end;
   beta = alpha\beta; errx = sum(abs(beta)); guess = guess + beta;
   if errx < tolx; break; end;
end
a = guess;
```

Some MATLAB Commands Commonly Used in Statics

`sin(x)`	compute the sine of the number x
`exp(x)`	compute the exponential of the number x
`cos(x)`	compute the cosine of the number x
`sqrt(x)`	compute the square root of the number x
`linspace(1,10,9)`	creates a vector of 9 elements evenly spaced, starting with 1 and ending with 10
`length(x)`	returns the length of the vector x
`size(A)`	returns the length of the vector x
`dot(x,y)`	computes the cross product of the two vectors x and y
`cross(x,y)`	computes the cross product of the two vectors x and y
`norm(x)`	computes the Euclidian norm of the vector x
`A'`	computes the transpose of the array A
`det(A)`	computes the determinant of the square matrix A
`inv(A)`	compute the inverse of the matrix A
`eig(A)`	compute the eigenvalues and eigenvectors of the matrix A
`syms ("x')`	declares that x is a symbolic variable
`diff(f)`	differentiates f symbolically or if f is a vector of numbers diff takes the difference between its elements
`solve(f-10)`	symbolically solves the equation f 10, f a symbolic expression
`int(f,s)`	symbolically integrates the function f with respect to s
`ezplot(f,[a,b])`	plots the symbolic function f between the numbers a and b
`subs(f,x,s)`	substitute x by s in the symbolic expression f
`compose(f,g)`	computes f(g(x)), f and g symbolic functions of x
`fzero('name',a)`	looks for a zero of the function stored in name near the numera.
`plot(t,x)`	plots x as a function of t where x and t are the same size vectors
`plot3(a,b,t)`	constructs a 3d plot of the vectors a nd b as functions of t
`quad8('name',a,b)`	computes the integral of the function stored in the name.m between the numbers a and b

INDEX